Environmental Remote Sensing

Environmental Remote Sensing

Editor: Fred Byron

R CALLISTO REFERENCE

www.callistoreference.com

Callisto Reference,
118-35 Queens Blvd., Suite 400,
Forest Hills, NY 11375, USA

Visit us on the World Wide Web at:
www.callistoreference.com

ISBN: 978-1-64116-184-8 (Hardback)

Cataloging-in-Publication Data

Environmental remote sensing / edited by Fred Byron.
     p. cm.
Includes bibliographical references and index.
ISBN 978-1-64116-184-8
1. Environmental monitoring--Remote sensing. 2. Remote sensing--Environmental aspects.
3. Earth sciences--Remote sensing. 4. Remote sensing. I. Byron, Fred.
QE33.2.R4 E58 2019
621.367 8--dc23

Table of Contents

Preface

The science and technology of environmental remote sensing refers to the use of aircraft- and satellite-based sensors to gather information about the Earth's atmosphere, oceans, glaciers, etc. It is different from on-site observation, as it does not include making physical contact with the observation site or the object of study. This technology is mostly used in conducting land surveys and guiding investigations in the fields of geology, ecology and hydrology. The different tools of environmental remote sensing are light detection and ranging (LIDAR), laser and radar altimeters, hyperspectral imagers, ultrasound and radar tide gauges, sonar, etc. This book is a valuable compilation of topics, ranging from the basic to the most complex advancements in the environmental applications of remote sensing. Also included herein is a detailed explanation of the various concepts and applications of environmental remote sensing. This book is a vital tool for all researching and studying this field.

After months of intensive research and writing, this book is the end result of all who devoted their time and efforts in the initiation and progress of this book. It will surely be a source of reference in enhancing the required knowledge of the new developments in the area. During the course of developing this book, certain measures such as accuracy, authenticity and research focused analytical studies were given preference in order to produce a comprehensive book in the area of study.

This book would not have been possible without the efforts of the authors and the publisher. I extend my sincere thanks to them. Secondly, I express my gratitude to my family and well-wishers. And most importantly, I thank my students for constantly expressing their willingness and curiosity in enhancing their knowledge in the field, which encourages me to take up further research projects for the advancement of the area.

Editor

Utilization of Ground-Penetrating Radar and Frequency Domain Electromagnetic for Investigation of Sewage Leaks

Goldshleger Naftaly and Basson Uri

Additional information is available at the end of the chapter

Abstract

Fact 1: Underground sewage pipe systems deteriorate over time, developing cracks and joint defects; therefore, leakage is inevitable. Fact 2: The massive worldwide urbanization process, together with rural development, has meaningfully increased the length of sewage pipelines. Result: The concomitant risk of sewage leaks exposes the surrounding land to potential contamination and environmental harm. It is therefore important to locate such leaks in a timely manner, enabling damage control. Advances in active remote-sensing technologies (GPR and FDEM: ground-penetration radar and frequency domain electromagnetic) were used to identify sewage leaks that might cause pollution and to identify minor spills before they cause widespread damage.

Keywords: Active remote sensing, FDEM, GPR, Sewage leak, Contamination, Water pollution

1. Introduction

Water pollution is the contamination of bodies of water such as aquifers, lakes, ponds, rivers and oceans. This contamination occurs due to direct or indirect discharge of pollutants into the water bodies, without a suitable treatment to remove harmful compounds (pollutants may simply be defined as substances added to the environment that do not belong there). A substantial proportion of water and environmental contaminants are due to leaks from underground sewage pipeline systems in rural, urban and industrial areas, since any sewage pipeline system deteriorates over time, developing cracks and joint defects. Therefore, if sewage pipeline systems are not maintained properly, it is only a matter of time before the sewage leaks out and contaminates the surrounding groundwater and surface water.

Here, we suggest detecting sewage leaks from pipeline systems using two orthogonal active remote-sensing methods: (I) ground-penetrating radar (GPR) and (II) frequency domain electromagnetic (FDEM). Our hypothesis is that GPR and FDEM screening, which creates subsurface images around and along pipeline systems, will enable the extraction of residual signals and the detection of meaningful leaks. Like most complex near-surface detection missions, detection of sewage leaks in an urban environment requires a professional understanding of the regional setting, from geomorphological, environmental and engineering perspectives.

Advances in remote-sensing technologies now enable their use to identify leakage that is potentially responsible for pollution and to identify minor spills before they can cause widespread damage. The detection of pollutants using GPR [1], was based on the research of Basson [2]. Basson [3] presented a combination of GPR and FDEM methods to detect and monitor saline contaminants in agricultural fields. Goldshleger [4, 5] demonstrated the ability to detect saline-affected soils using remote-sensing methods, toward improved management of these soils. Basson [6] described the detection of subsurface water/sewage/drainage pipe systems and leaks/contamination from such pipes. Ben-Dor [7, 8] reviewed remote-sensing–based methods to assess soil salinity and improve the management of salinity-affected soils. Ly and Chui [9] developed accurate representations of weep holes and leaky sewage pipes, and further showed the systems' long-term and short-term responses to rainfall events. Their simulation results provided a better understanding of local-scale migration of sewage leaks from a sewage pipe to nearby storm water drains. The last few years in Israel have seen increasing use of new methods based on active remote-sensing tools to study subsoil quality. These tools include GPR and underground monitoring systems measuring spatial moisture content, such as FDEM in the subsurface. The use of GPR is based on a method that was originally developed for measuring sand dunes of medium moisture content at an unsaturated resolution of a few percentage points [2]. The GPR helped define the possible reason for emerging high-salinity areas, such as a subsurface regional structure that reduces water infiltration into the deeper groundwater position [5]. The FDEM method provided a very important view of salt contamination in the soil layers (except the root zone layer) and also pinpointed areas with salinity problems. The images obtained from FDEM readings provided a subsurface view that also helped identify the reason for the high salinity in certain areas. In the soil salinity experiment in Israel, a severe defect in the drainage pipelines could be observed, which helped the farmers solve the problem before the subsequent season [5].

The present study focuses on the development of these electromagnetic (EM) methods to replace conventional acoustic methods for the identification of sewage pipe leaks. EM methods provide an additional advantage in that they allow mapping the fluid transport system in the subsurface. Leak-detection systems using GPR and FDEM are not limited to large amounts of water, but can also detect leaks of tens of liters per hour, because they can locate increases in pipes' or tanks' environmental moisture content that amount to only a few percentage points. The importance and uniqueness of this research lies in the development of practical tools to provide a snapshot of the spatial changes in soil moisture content to depths of about 3–4 m (in areas with asphalt overlay) at relatively low cost, in real time or close to real time. Spatial measurements performed using GPR and FDEM systems allow monitoring many tens of thousands of measurement points per hectare, thus providing a picture of the spatial situation along the pipelines. The main purpose of this study was to develop a method for detecting sewage leaks using the above-proposed geophysical methods, as the resultant contaminants can severely affect public health. We focused on identifying, locating and characterizing such leaks in sewage pipes in residential and industrial areas.

2. Methods

In recent years, there has been an increase in the use of active remote-sensing tools, such as GPR (Figure 1a) and subsurface FDEM (Figure 1b), for measuring the subsurface's EM velocity and dielectric constant (GPR), and its electrical conductivity profile and magnetic susceptibility (FDEM).

(a)

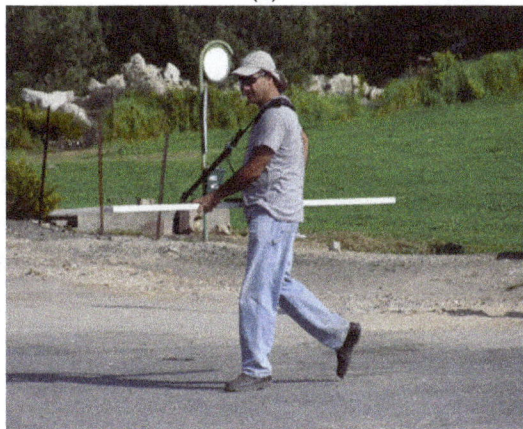

(b)

Figure 1. Taking measurements with the RAMAC GPR (a) and Gem-2 FDEM (b) in the study area.

Passive remote-sensing spectroscopy of ground surface and cross-sections using an optical fiber termed SPSP (subsurface-penetrating spectral probe), developed [10] and have been conducted as well. This study focuses on remote-sensing tools to replace acoustic methods [11, 12, 13]. EM methods provide the added advantage of being able to map underground liquid-carrying pipelines. Ground leak-detection systems using GPR and FDEM are not limited to large amounts of water: small leaks of tens of liters per hour can be detected in the environment by comparing medium-dry to minimum moisture content in the pipeline and the canal zone.

Our aim was to develop practical tools that would provide a snapshot of changes in spatial soil moisture content to depths of about 3–4 m in areas covered with asphalt at relatively low cost and in real time. The spatial measurements were performed with FDEM and GPR systems that allow measuring tens of thousands of points per hectare and thus enable monitoring the spatial situation along the pipeline.

2.1. FDEM

Traditionally, the electrical method "measures" apparent resistivity using electrodes that require ground contact in a DC electrical survey, while the EM method "measures" apparent conductivity without ground contact. The EM method, known as a "potential method", involves transmitting and receiving EM fields, commonly using a set of coils. The common unit of resistivity is ohm-m and conductivity is its inverse, in Siemen/m. The apparent resistivity ρ_a is defined in DC resistivity as:

$$\rho_a = 2\pi G \frac{\Delta V}{I} \tag{1}$$

where ΔV is the voltage between a pair of potential electrodes, I is the current that flows through another pair of source electrodes, and G is the geometric factor that depends on the geometry of the electrodes. For a Wenner array that uses four equally spaced electrodes, for instance, G is the electrode spacing itself. Even for this simple array, each electrode spacing generates a different apparent resistivity because the spacing controls the volume of the subsurface sampled by the measurement. It is only when the earth is a homogeneous half space that the apparent resistivity is the same as the true resistivity.

Similarly, apparent conductivity is only same as the true conductivity when the earth is a homogeneous half space. As an example, consider a pair of horizontal coils separated by a distance r. A routinely measured quantity is called the *mutual coupling ratio* which, for a horizontal coplanar (or vertical dipole) coil configuration over a layered earth as derived by [14, 15, 16, 17], among others is written as:

$$Q = \frac{Hs}{Hp} = -r^3 \int_0^\infty \lambda^2 R(\lambda) J_0(\lambda r) e^{-\lambda h} d\lambda \tag{2}$$

Hp and Hs are the primary and secondary fields at the receiver coil; $J0$ is the 0th order Bessel function; r is the coil separation and h is the sensor height above the ground. Q represents the secondary field normalized against the primary field at the receiver coil. Most frequency-domain sensors measure Q in parts per million (ppm). The kernel R corresponding to a homogeneous half space is:

$$R(\lambda) = \frac{\lambda - \sqrt{\lambda^2 + \imath 2\pi f \mu \sigma}}{\lambda + \sqrt{\lambda^2 + \imath 2\pi f \mu \sigma}} \qquad (3)$$

where f is the transmitter frequency in Hz, μ the magnetic permeability and σ the half-space conductivity. Based on Q measured at a particular frequency over a real (heterogeneous) earth, we can invert Equation (2) to obtain the *apparent* half-space conductivity σ_a. It is obvious from Equation (2) that the resulting σ depends on coil separation, sensor height and frequency. In addition, each coil configuration (vertical coplanar, coaxial, etc.) has a different formula for Q. Figure 2 shows a coplanar coil pair at height h above layered earth [18], and a damped least-squares inversion based on singular value decomposition to solve the nonlinear inverse problem.

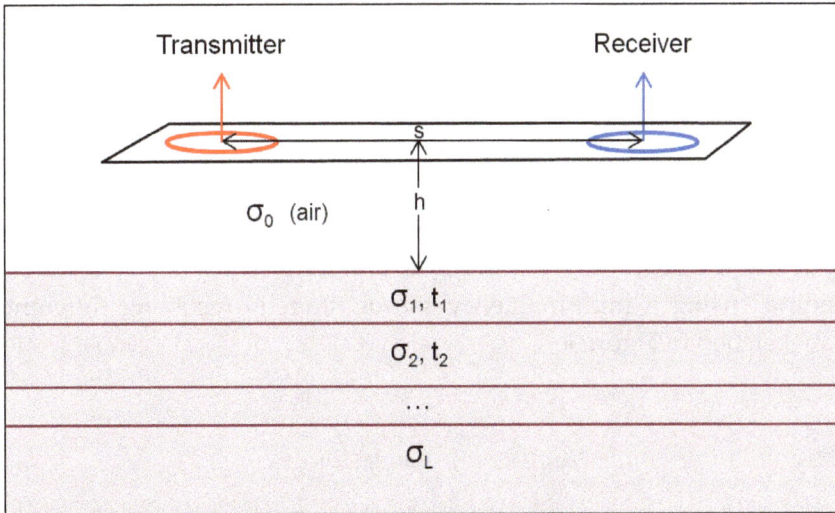

Figure 2. Geometry of the horizontal coplanar electromagnetic sensor over layered earth where σ is the conductivity, t is the thickness of each layer, the subscripts stand for the number of layers, s is the coil separation and h is the sensor height [18].

Figure 3 shows the responses of the Gem-2 sensor over a half space as a function of induction number:

$$\theta = (\sigma \mu \omega / 2)^{1/2} s \qquad (4)$$

where ω is the angular frequency, μ is the magnetic permeability and s is coil separation.

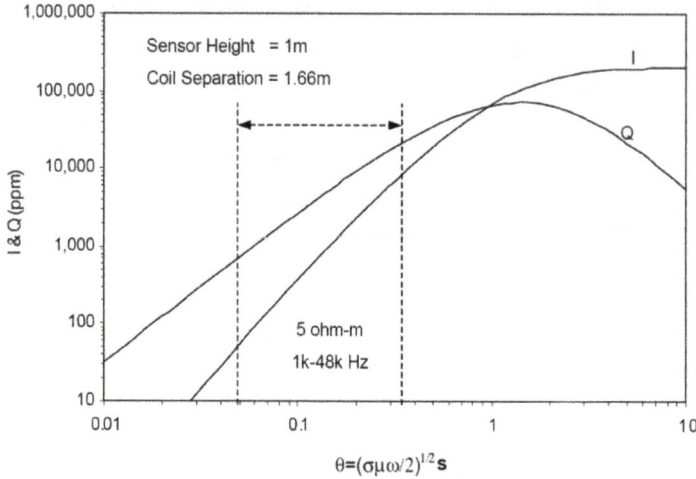

Figure 3. The in-phase and quadrate responses as a function of induction number (from Huang and Won, 2003).

The depth of investigation of an EM system can be estimated using the skin depth δ, which is defined in classical EM theory as the distance in a homogeneous medium over which the amplitude of a plane wave is attenuated by a factor of $1/e$, or about 37% of its original amplitude. The skin depth δ is:

$$\delta = \sqrt{\frac{2}{\sigma\mu\omega}} \tag{5}$$

The skin depth and the ability to transmit in several frequencies allows us to perform "frequency sounding" using a multifrequency sensor, thereby resolving different depths of penetration as sketched in Figure 4.

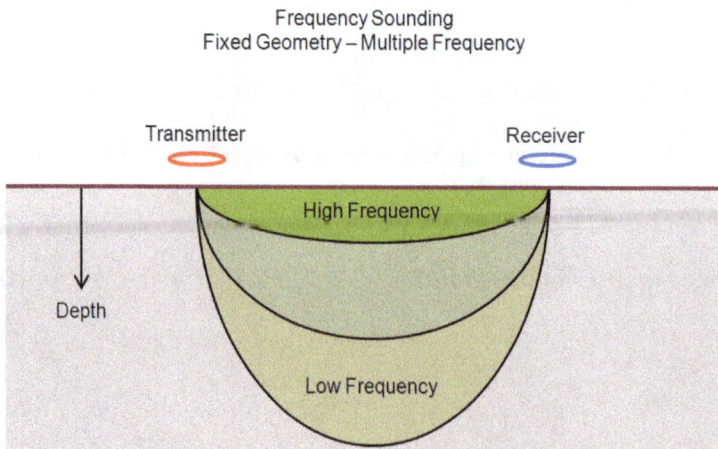

Figure 4. Frequency sounding for various depths using a multifrequency FDEM sensor such as Gem-2.

2.2. GPR

GPR, a reflection-scattering imaging method, is widely used for subsurface imaging in geophysics. GPR uses high frequencies (wavelengths; MHz–GHz). EM waves may form images of the subsurface by transmitting radar pulses into the ground and receiving the deflected waves from the interfaces below. Using wave methods and analysis, GPR images can be analyzed for their derived electrical properties and subsurface characteristics and for spatial mapping of water content [2, 3]. The range resolution is a function of the subsurface dielectric constants and the wave's frequency. It may vary from several centimeters to several tens of centimeters at the relevant effective frequencies [19, 20] For a certain wavelength, the penetration of GPR waves into the subsurface is mainly a function of the host material's conductivity, and therefore GPR waves decay significantly in conductive and saline soils. Using wave methods and analysis, GPR images can be analyzed for their derived electrical properties and subsurface characteristics and for spatial mapping of water content [2,3], as described in the following model.

The connection between the EM velocity and dielectric constant is expressed as:

$$v = \frac{c}{\sqrt{k}} \tag{6}$$

where c is the speed of light in a vacuum and k is the dielectric constant.

The dielectric constant of water (k_w) is about 80. The dielectric constant of air (k_a) is 1. The dielectric constant of common "dry" soil ($k_{dry\ soil}$) with residual moisture content can range between 6 and 15 (the effective dielectric constant of dry soil is determined according to volumetric mixing ratios between soil, water and air components).

The difference in the effective dielectric constant of "dry" and "wet" soils is mainly a function of the ratio between the air and water volumes, when the volumes are normalized to:

$$V_{dry\ soil} + V_w + V_a = 1 \tag{7}$$

then:

$$k_{eff} = k_{dry\ soil} V_{dry\ soil} + k_w V_w + k\left(1 - V_w\right) \tag{8}$$

The maximal soil–water absorbency is a strong function of the effective porosity.

3. Leak detection in Ariel

Ariel is a small city (about 20,000 residents) in Israel, located in the central highland region known as the Samarian Hills. It is situated 40 km (25 miles) east of Tel Aviv and 40 km west of the Jordan River. It is situated 700 m (more than 2000 feet) above sea level. The city stretches over 12 km (8 miles) in length and 2 km in width. The research was performed with Yuvalim, the company that is responsible for maintaining the water and sewage network in the Ariel area and for supplying available water to residents. The mutual research was performed to identify sewage leaks before they pollute and damage the surrounding area. The research was supported by the Israeli Water Authority. The work was performed in several stages.

3.1. Selecting study sites

Areas were selected in Ariel for system calibration (Figure 5). Two areas were chosen for the method calibration: the first was an industrial area and the second a residential area, both with well-mapped networks of water and sewage pipes. These areas were selected on the basis of information from computerized data, observations, field visits, use of orthophotos, aerial photography and geological and pedological data.

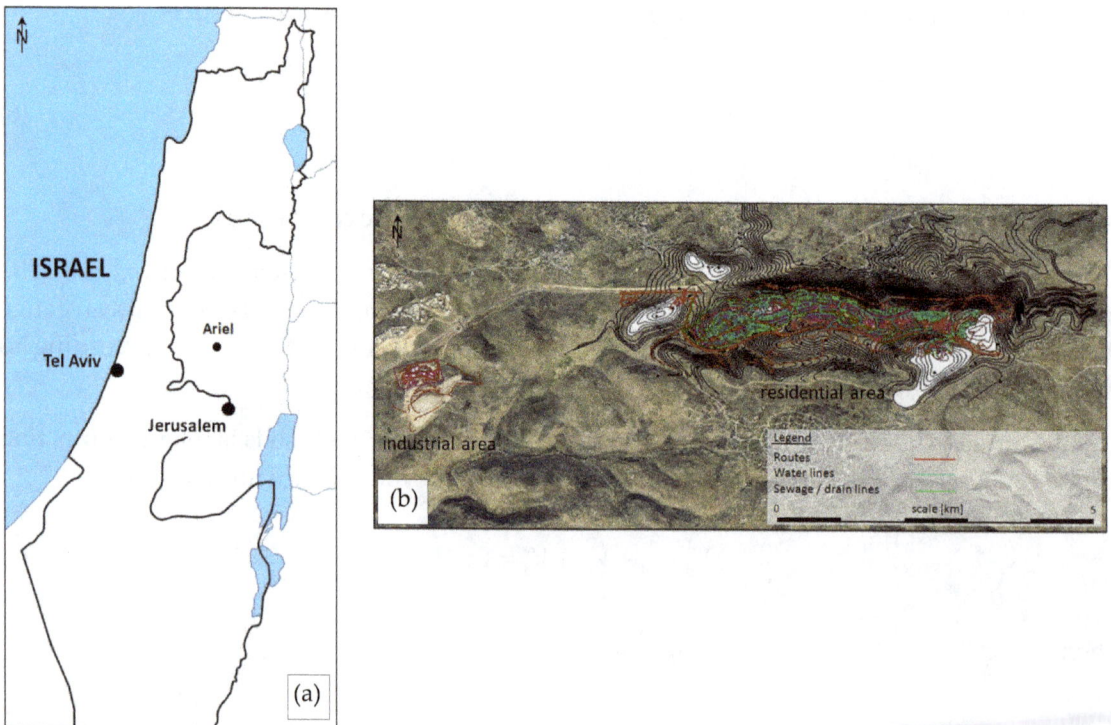

Figure 5. Maps showing Ariel's location (a) and the drainage infrastructure, sewerage and water supply for this city (b).

3.2. Soil characterization

To characterize the pedological structure of the subsurface layers, excavations were performed. We sampled grain size, void content and porosity, moisture content, soil density and soil characteristics. We dug a channel in an underground sewage pipe replacement area at the experimental sites. Figure 6 presents the characterization of the sub layer.

Figure 6. Soil subsurface cross-section at site 1. Wooden pegs mark the changing soil layers.

The soil in the area is red Mediterranean, also known as Terra Rossa [21] and Lithic ruptic Xerochrept [22]. Terra Rossa occurs in areas where heavy rainfall dissolves carbon from the parent calcium carbonate rock and silicates are leached out of the soil, leaving residual deposits that are rich in iron hydroxides, causing the red color. Such areas are usually depressions within limestone. The soil was sampled in a 0.5-m-wide ditch at a depth of 2 m. The area has an easterly aspect, with an average elevation of 400 m above sea level. The local slopes vary between 7% and 25%. Soil texture was clay loam with an average composition of 45% sand, 25% silt and 30% clay. The sand content increased toward the lower part of the area. The average lime content was 30%. Rock fragments of up to 40 cm appeared together with the soil.

3.3. GPR calibration

Calibration of the GPR system to the subsurface properties of the cross-section in a dry state (without leakage) is shown in Figure 7. The depth to the pipe was measured in a nearby manhole.

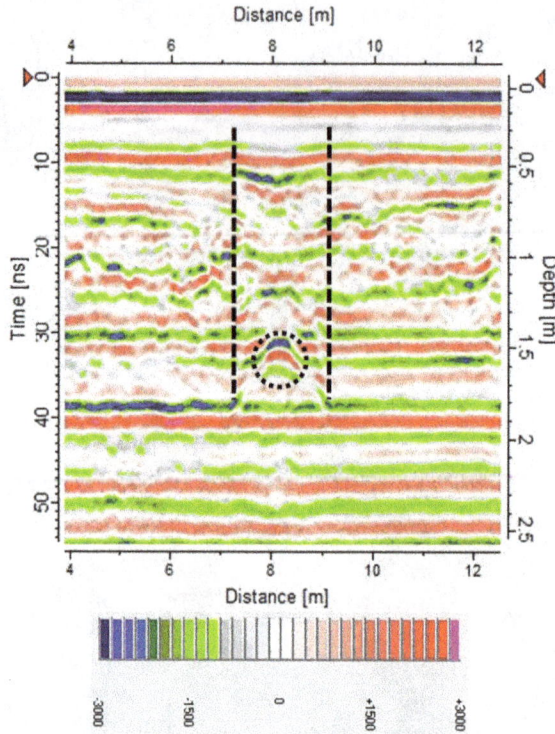

Figure 7. Part of the GPR profile performed for calibration of the GPR system in the Ariel industrial zone, on the road close to a rubber factory. The black circle displays diffraction created by the drain pipe. Above it, the trench is detected as well. The horizontal scale describes the measurement location (in meters) along the profile. The vertical scales describe the time (in nanoseconds) and depth (in meters). The amplitude–intensity scale is shown as well.

Figure 7 shows the results of advanced processing of a cross-section for calibration of the system in the industrial area. On the horizontal scale, simulations are described above the measurement location along the incision in meters; the vertical scales describe the time and depth of the reflections on a timescale of 50 ns and scale depth of 2.5 m below the surface (the strength of the reflections is graded according to the color scale in Figure 7, where the diffraction created by the drainage pipe can be deduced from a return time from the pipe of approximately 32 ns). The diffraction depth is 2.45 m, and the data from the system matches the data measured on the ground. This adaptation makes it possible to determine the velocity of the EM wave. The average measured subsurface speed of the EM wave (v) was 0.093 ± 0.001 m/ns at the Ariel industrial site. It is important to note that the speed of the wave depends on the directly calculated form and location of the anomaly and thus data processing is critical to the research results.

3.4. The experimental site

The experimental site for sewage pipeline and manhole leaks was located near Ariel's old stadium, not far from HaAtsmaut Street (Figure 8, blue rectangle), where a project for the replacement of old sewer pipes has been initiated.

Figure 8. The experimental site is located at the western end of the sewage line adjacent to HaAtsmaut Street (blue rectangle). It includes 12-in. diameter iron pipes carrying on the order of 1000–1200 m^3 sewage water per day, and an average 100 m^3/h during peak flow.

Leakage was initiated in two places at the western site by cracking the sewer pipes close to their bottom side. One crack was made about 6 m from the sewage pit in the northern iron pipe using an electrical disk that created a wedge-shaped hole 15–20 cm in diameter; the second crack was also a circle of 15–20 cm diameter in the lower part of the pipe (Figure 9). The experimental site was monitored daily by radar and FDEM before the start of and during the controlled leakage.

controlled leaks

Figure 9. Pictures of the two cracks made in the sewer pipes for the controlled leakage experiment.

4. Results

Daily monitoring with the FDEM method included five cross-sections: four were parallel to the sewer pipeline and the fifth was above it, running on each side of the pipeline at a distance of 0.5 m. During the experiment, FDEM scanning was performed to qualify the effect of moisture on the soil cross-section. Figure 10 shows the status of the subsurface before the start of the controlled leak; it was in a relatively dry state characteristic of the month of May at this site.

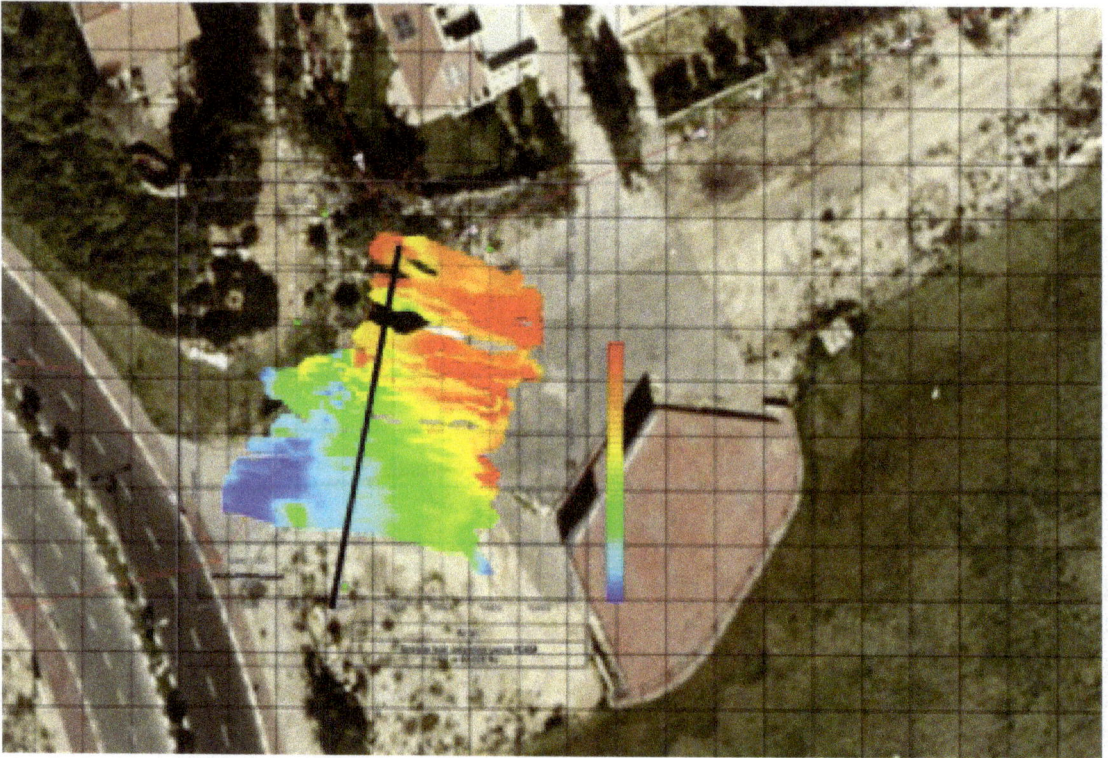

Figure 10. Map of the integrated electrical conductivity at 60,025 Hz before the start of the controlled leak at the western site (locations of the measurements are shown by the blue rectangle in Figure 8). The map is based on measurements performed with a GEM-2 FDEM sensor. The location of the sewer pipe is marked with a black line. Data were collected prior to the leak with dimensional scanning of approximately 30 m × 25 m. Lower conductivity (σ) values (11 mS/m) appear in blue-green in the southwestern corner of the area, while the highest conductivity appears in red-purple (55 mS/m) in the northeastern part of the map. These conductivity changes suggest anomalous subsurface moisture from the water pipe near the old stadium, as well as the accumulation of water from the slope, where there is a garden.

Figure 11 shows a pronounced increase in electrical conductivity of about 40 mS/m after 4 days of controlled leakage. The area has high conductivity because of changes in wetness due to a significant increase in liquid as a result of the sewage flow.

The results of the FDEM measurements conducted 10 days after the beginning of the controlled leak are presented in Figure 12. This picture may look similar to Figure 11 in terms of colors, but their intensity has increased due to an increase in the conductivity values to about 152 mS/m.

On the map in Figure 12, low visibility, reflecting low electrical conductivity, is shown in blue-green shades, high visibility in red-colored shades. Purple indicates sewer leakage on the background of the driest area, highlighting the differences in moisture. A wide area can be seen west of the pipe (black line in Figure 12) with relatively low electrical conductivity compared to the rest of the region. Northeast of the pipe, there is high electrical conductivity resulting from the spillover of sewage water.

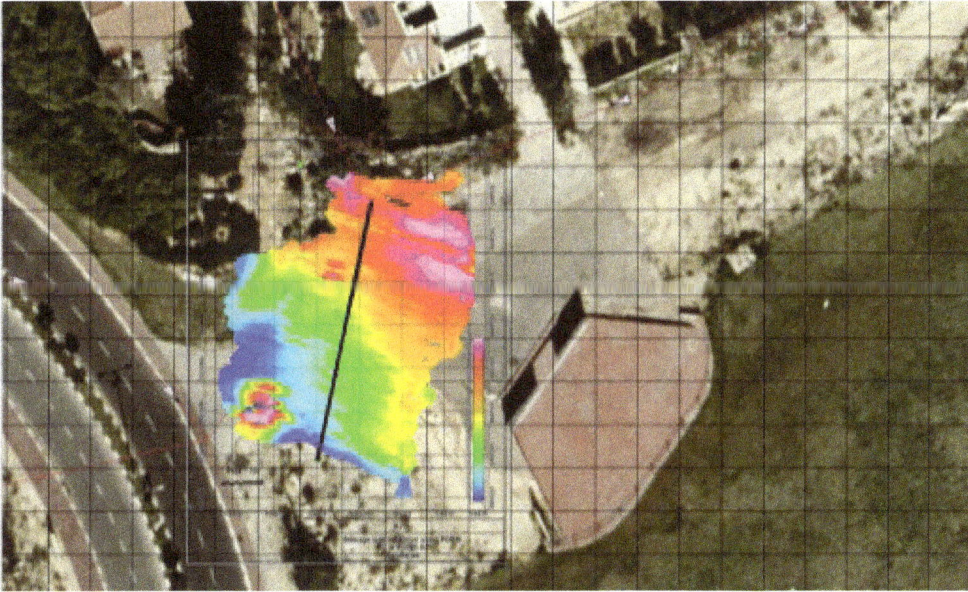

Figure 11. Map of the electrical conductivity at 60,025 Hz after about 4 days of leakage. Measurements were collected during the sewage leak, under wet conditions, with the GEM-2 sensor (locations of the measurements are shown by the blue rectangle in Figure 8). The location of the sewer pipe is marked with a black line. The highest conductivity value was about 95 mS/m. The significant increase in electrical conductivity is a result of the sewer liquids that were spilled during the 4 days of the controlled leak, in both the southwestern and northeastern sides of the area, probably due to a subsurface topography gradient.

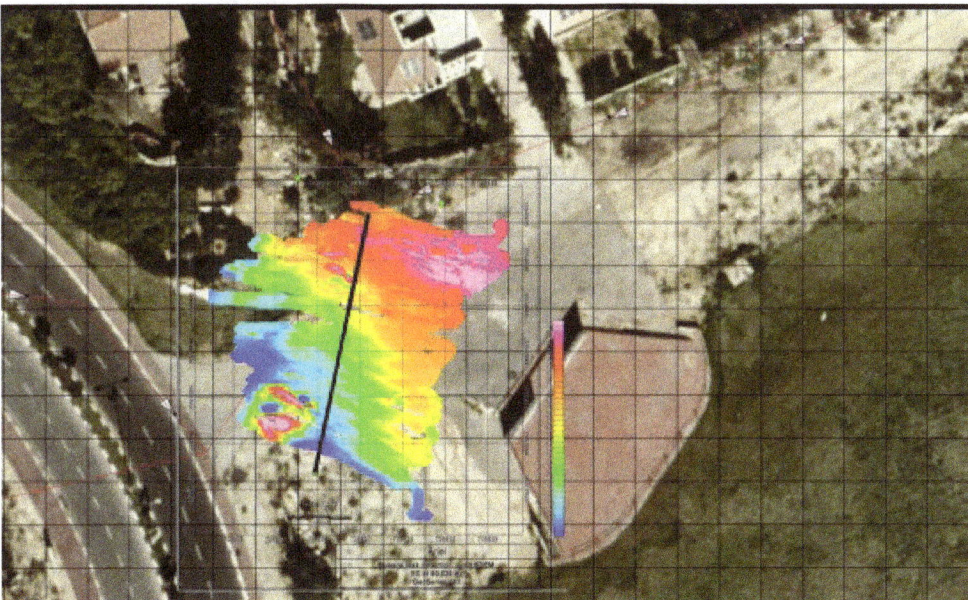

Figure 12. Map of integrated electrical conductivity at 60,025 Hz. Measurements were collected with the FDEM system, under wet conditions, after 10 days of controlled leakage (locations of the measurements are shown by the blue rectangle in Figure 8). Electrical conductivity ranged from 0 to 152 mS/m. Low conductivity is expressed in blue-green shades, high conductivity in purple-red colors.

Figure 13 shows maps made by FDEM monitoring of electrical conductivity at various frequencies in the first tested area. The maps are arranged, from left to right, at increasing frequencies and depth: the frequencies were 2,025 Hz, 4,725 Hz, 11,025 Hz, 25,725 Hz and 60,025 Hz, each frequency representing a 30 cm increase in depth. The low-visibility electrical conductivity is represented by blue-green hues, and the high-visibility electrical conductivity by red-purple hues. There were a few quantitative differences in the map scales.

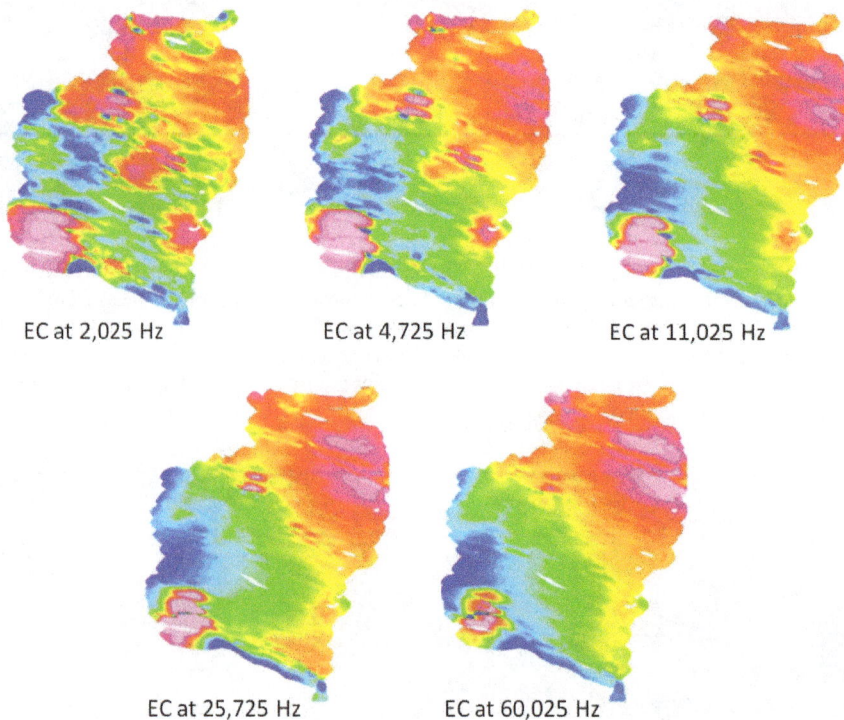

Figure 13. Maps made by FDEM monitoring of electrical conductivity at 2,025 Hz, 4,725 Hz, 11,025 Hz, 25,725 Hz and 60,025 Hz. The lower EC values are represented by blue-green hues, and the higher EC values by red-purple hues. There were a few quantitative differences between the maps' scales.

Four sections, two on each side of the sewer, were monitored by GPR and are shown in Figure 14. The distance between the main radar cross sectional cuts was approximately 0.5 m. The radar sections shown in Figure 14 were collected with an antenna at a nominal frequency of 250 MHz over the location of the underground sewage pipe at the first (western) test site. The first cross-section was obtained before the leak started and reflects the typical dry state of the ground in May. An incision was made a few days after the initiation of the leak and shows a relatively wet subsoil. The right cross-section shows an incision made at a lower depth, 10 days after leak initiation, indicating a further increase in wetness. Similar data processing was carried out for the three cross-sections to highlight their differences.

Dry section

Wet section

Max. Wet section

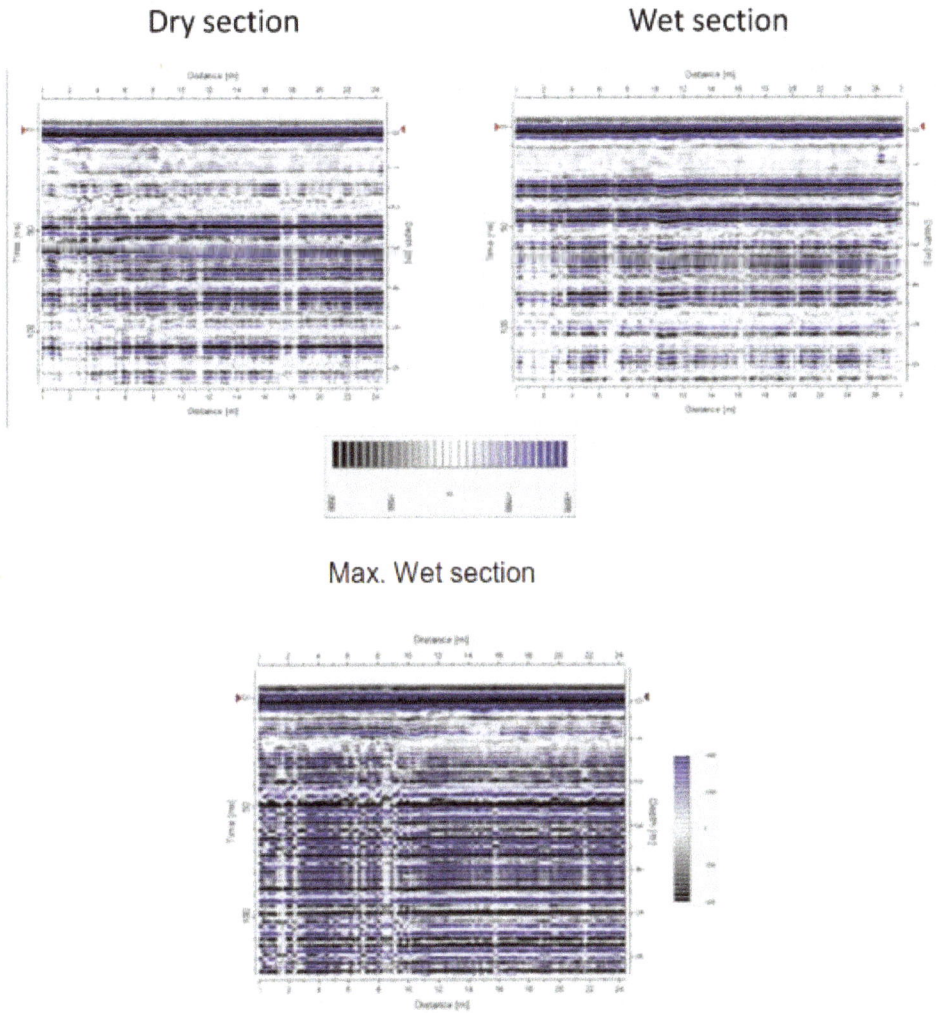

Figure 14. Soil moisture reflected by GPR cross-section (locations of the measurements are shown by the blue rectangle in Figure 8).

5. Modeling subsurface moisture content

Moisture content was computed on the basis of subsurface GPR and FDEM measurements and its spatial spread was obtained for calibration and wetness testing with water- and sewage-carrying pipelines. In these experiments, radar velocities were measured and dielectric constants were computed. Their correlations were used to measure the moisture content from data collected in the residential and industrial neighborhoods.

The computation of moisture content using GPR was based on the method developed by Basson [2] From the calibration measurements conducted at the end of May 2012, the average subsurface EM wave velocity was 0.093 ± 0.001 m/ns. The calculated dielectric constant during this period was about 10.4. This value is low but not minimal, as minimal moisture content is typically found in the mid-to-late summer months (according to data from the Israel Meteorological Service, the rain that accumulated in the area in the months before the GPR measurements amounted to about 161 mm).

The velocity of EM waves in a substance is mainly a function of that substance's bulk dielectric properties and moisture content. When a substance is composed of a mixture of materials, the velocity is a function of their mixing ratios. In the case of a subsurface environment, we can treat the substance as a bulk property composed of soil, rock, minerals and organic materials mixed with air and water. When the rate of air increases, the velocity increases as well. However, when the moisture content increases, the average dielectric constant decreases as well and fro equation (6) it can be seen that the EM velocity (v) decreases as well.

The difference in the effective dielectric constant of "dry" and "wet" soils is mainly a function of the ratio between the air and water volumes, when the volumes are normalized according to equations (7) and (8). The maximal soil–water absorbency is a strong function of the effective porosity. For soils in the Ariel region, the effective porosity can vary from 40% to 60%. We used an average effective porosity of 50% in our computations. Therefore, the possible mixing ratios relative to the normalized volume are:

$$V_{\text{dry soil}} = 0.5 V_{\text{tot}} \tag{9}$$

$$V_{\text{w}} + V_{\text{a}} = 0.5 V_{\text{tot}} \tag{10}$$

Since $k_{\text{dry soil}}$ is the effective dielectric constant measured using GPR imaging for a soil with residual moisture content and since $k_{\text{a}} = 1$:

$$k_{\text{top soil}} = 0.5 k_{\text{dry soil}} + 80 V_{\text{w}} + 0.5 - V_{\text{w}} \tag{11}$$

The radar wave velocity for "dry" soil at the surface will be measured and is expected to vary with the GPR and its value, $v_{\text{top soil}} \sim 0.07\text{--}0.14$ m/ns. From Equation (1), this velocity range can reflect dielectric constant values of ~4.6–18.4 for $k_{\text{top soil}}$. For example, for maximal dielectric constant values of 5–14 for delicate quartz-based soils and for the presented computations, the moisture content in the surface can vary as $V_{\text{w top soil}} \sim 0.4\%\text{--}2.1\%$. In the same way, we can investigate deeper soils where the moisture content is expected to be greater. The average radar wave velocity ($v_{\text{humid soil}}$) measured by the GPR at the calibration site in Ariel at the end of May is 0.093 m/ns. Using Equation (1), this velocity reflects a dielectric constant value ($k_{\text{humid soil}}$) of 10.406. The additional volume of water needed to increase the dielectric constant from 4.6 to 10.41 can be computed as:

$$\Delta k = 5.81 \tag{12}$$

$$\Delta k = 80\Delta V_w \tag{13}$$

$$\Delta V_w = 7.26 \tag{14}$$

We develop a moisture content model using relative values of the moisture content (based on Equations (6–14)) causing an increase in electrical conductivity as measured by the FDEM. We had to consider the overall subsurface features, such as texture, density and effective porosity, as well as the content of salts in soils irrigated with brackish effluent water. The model results are presented in the graph in Figure 15.

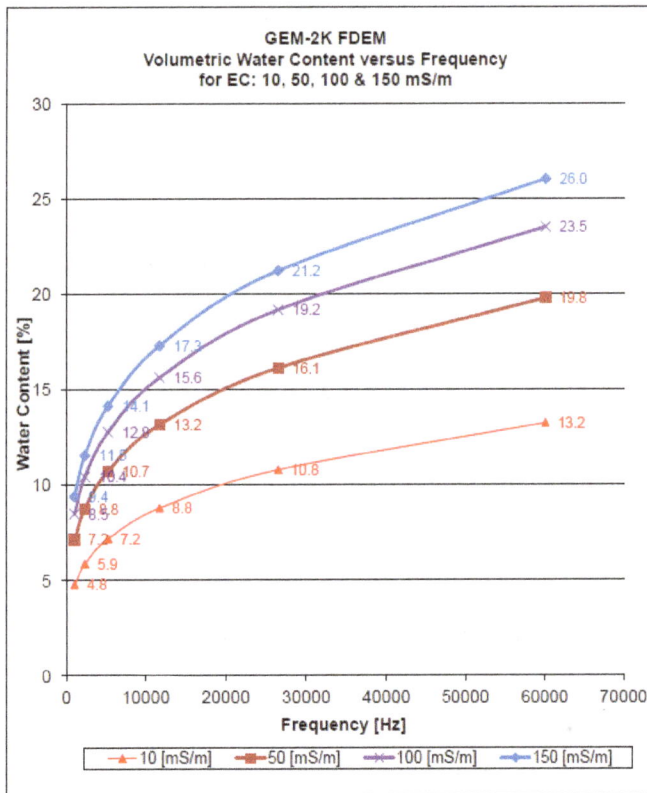

Figure 15. Volumetric moisture content calculated from measurements and from the FDEM model in the experimental zones in Ariel (accuracy ±10% of the measured value).

6. Discussion and conclusions

We introduced a combination of GPR and FDEM orthogonal methods to detect subsurface leaks from a sewage pipeline system. The rationale for this combination is to increase the probability of detection, especially in complex urban environments and when the soil–rock setting can vary from relatively resistive to relatively conductive. The results of our study indicate that even minor leaks, such as the minor controlled leaks created in the experiment, and changes in the subsurface moisture content can be accurately detected. We could detect sewage leakage, as well as its progress. The combination of the two methods enabled not only the detection of the leak but also a qualitative assessment of its size. Factors affecting the ability to detect leaks were limited by the soil–rock conductivity, as well as the density of the terrain and subterrain systems and structures. The geophysical methods may detect sewage effluent flow paths as well as the contaminant in the soil.

The limestone and dolomite bedrock in the Ariel area is suitable for GPR mapping. The clarity of the GPR profile enabled analysis and interpretation of the physical data with good accuracy. We could detect sewage leakage, as well as its progress. The anomalous moisture of the leakage accumulating around the sewage pit in the southwest research area validated the efficiency of the methods.

Acknowledgements

The research was supported by the Israel water authority and by the Water Cooperation of Yuvalim. We would like to thank Mr. Omer Shamir from GeoSense for assistance with the data collection.

Author details

Goldshleger Naftaly[1,2]* and Basson Uri[3]

*Address all correspondence to: Goldshleger1@gmail.com

1 Civil Engineering Ariel University, Israel

2 Israel Ministry of Agriculture, Beit Dagan, Israel

3 Geosense Ltd., Even Yehuda, Israel

References

[1] Basson, U., and Ben-Avraham, Z., 1994. Subsurface spatial mapping of pollutants concentrations using ground penetrating radar. Proceedings of the 25th Annual Meeting of the Israel Society for Ecology and Environmental Quality Sciences, Tel-Aviv, Israel, 3–4 May 1994, p. 29.

[2] Basson, U., 1992. Mapping of Moisture Content and Structure of Unsaturated Sand Layers with Ground Penetrating Radar. M.Sc. thesis.Tel-Aviv University, Tel-Aviv, Israel.

[3] Basson, U., 2000. Imaging of Active Fault Zone in the Dead Sea Rift: Evrona Fault Zone as a Case Study. Ph.D. thesis. Tel-Aviv University, Tel-Aviv, Israel.

[4] Goldshleger, N., Mirlas, V., Ben-Dor, E., Eshel, M., and Basson, U., 2007. *Using Remote Sensing Methods for Improving the Management of Saline Affected Soils* ERSA, Conference Paris, France.

[5] Goldshleger, N., Livene, I., Chudnovsky, A., and Ben-Dor, E., 2012. Integrating passive and active remote sensing methods to assess soil salinity: a case study from Jezre'el Valley, Israel, Soil science 177(6), 392–401.

[6] Basson, U., 2007. Imaging and mapping subsurface infrastructures and buildings using GPR and FDEM electromagnetic methods. *Journal of Nondestructive News, Vol. 10, pp. 29-30.*

[7] Ben-Dor, E., Goldshleger, N., Eshel, M., Mirablis, V., and Basson, U., 2008.Combined active and passive remote sensing methods for assessing soil salinity. In: *Remote Sensing of Soil Salinization. Impact and Land Managemen* (G. Metternicht and A. Zinck, eds.), pp. 235–255. CRC Press, USA.

[8] Ben-Dor, E., Metternicht, G., Goldshleger, N., Eshel, M., Mirablis, V., and Basson, U., 2008. Review of remote sensing based methods to assess soil salinity. In: *Remote Sensing of Soil Salinization. Impact and Land Management* (G. Metternich and A. Zinck, eds.), pp. 39–60. CRC Press, USA.

[9] Ly, D.K., and Chui, T.F., 2012. Modeling sewage leakage to surrounding groundwater and storm water drains. *Water Science Technology* 66(12), 2659–2665.

[10] Ben-Dor, E., Heller, D. and A. Chudnovsky, 2008. 10. A novel method of classifying soil A A novel method of classifying soil profiles in the field using optical means Soil Science Society of American Journal, *72:1-13.*

[11] Klein, W.R., 1993. Acoustic leak detection. *American Society of Mechanical Engineers, Petroleum Division* 55, 57–61.

[12] Hough, J.E., 1988, Leak testing of pipelines uses pressure and acoustic velocity. *Oil and Gas Journal* 86, 35–41.

[13] Hunaidi, O., and Wang, A., 2004. Acoustic methods for locating leaks in municipal water pipe networks. *International Water Demand Management Conference, Dead Sea, Jordan, 30 May–3 Jun 2004.*

[14] Frischknecht, F.C., 1967. Fields about an oscillating magnetic dipole over a two-layer earth and application to ground and airborne electromagnetic surveys. *Quarterly of the Colorado School of Mines* 62, 326.

[15] Ward, S.H., 1967. Electromagnetic theory for geophysical applications. In: *Mining Geophysics* (S.H. Ward, ed.), pp. 13–196. Society of Exploration Geophysicists, Theory, USGS.

[16] Ward, S.H., and Hohmann, G.W., 1988. Electromagnetic theory for geophysical applications. In: *Electromagnetic Methods in Applied Geophysics* (M.N. Nabighian, ed.), pp. 130–311. Society of Exploration Geophysics, Theory, Tulsa, Oklahoma.

[17] Won, I.J., and Huang, H., 2004. Magnetometers and electro magnetometers. *The Leading Edge* 23(5), 448–451.

[18] Huang, H., and Won, I.J., 2003. Real-time resistivity sounding using hand-held electromagnetic sensor. *Geophysics* 68(4), 1224–1231.

[19] Davis, J.L., and Annan A.P., 1986. High resolution sounding using ground probing radar. *Geoscience Canada* 3: 205–208.

[20] Davis, J.L., and Annan, A.P., 1989. Ground penetrating radar for high resolution mapping of soil and rock stratigraphy. *Geophysical Prospecting* 37: 531–551.

[21] Dan, Y., and Raz, Z., 1970. *Soil Association Map of Israel*. Volcani Institute for Agriculture Research, Israel (in Hebrew).

[22] Soil Survey Staff, 1975. *Soil Taxonomy: A Basic System of Soil Classification for Making and Interpreting Soil Surveys*. US Department of Agriculture, Handbook 436, pp. 754.

2

Processing of Multichannel Remote-Sensing Images with Prediction of Performance Parameters

Benoit Vozel, Oleksiy Rubel, Alexander Zemliachenko, Sergey Abramov, Sergey Krivenko, Ruslan Kozhemiakin, Vladimir Lukin and Kacem Chehdi

Additional information is available at the end of the chapter

Abstract

In processing of multichannel remote sensing data, there is a need in automation of basic operations as filtering and compression. Automation presumes undertaking a decision on expedience of image filtering. Automation also deals with obtaining of information based on which certain decisions can be undertaken or parameters of processing algorithms can be chosen. For the considered operations of denoising and lossy compression, it is shown that their basic performance characteristics can be quite easily predicted based on easily calculated local statistics in discrete cosine transform (DCT) domain. The described methodology of prediction is shown to be general and applicable to different types of noise under condition that its basic characteristics are known in advance or pre-estimated accurately.

Keywords: Multichannel remote sensing data, automatic processing, denoising, lossy compression, performance prediction, DCT

1. Introduction

Remote-sensing (RS) data are widely used for numerous applications [1], [2]. Primary RS images acquired onboard of airborne or spaceborne carriers and intended for Earth surface monitoring are usually not ready for direct use and, thus, are subject to a certain preprocessing. This preprocessing can be carried out in several stages and includes the following operations: geo-referencing and calibration, blind estimation of noise/distortion characteristics, pre-filtering, lossless or lossy compression, [1], [2], etc. These operations can be distributed between onboard and on-land computer means (processors) in different ways depending upon many factors [3-5].

Regardless of the distribution of functions, the operations onboard are usually performed in a fully automatic manner (although there can be some changes in algorithm parameters by command passed from Earth). In turn, the operations carried out on land can be, in general, performed in an interactive manner and labor of highly qualified experts is exploited for this purpose. However, a certain degree of automation of on-land data processing is required as well. The need in processing automation is especially high if one deals with multichannel (e.g., hyperspectral) RS data [6], where the number of channels (components, sub-bands) can reach hundreds. Such RS images have become popular and widespread (available) currently due to their (potential) ability to provide rich information for various applications [6], [7].

Meanwhile, the multichannel nature of RS data results in new problems in their processing [3], [8]. The main problems and actual questions are the following:

- How to manage large volumes of acquired data with maximal or appropriate efficiency (here, different criteria of efficiency can be used)?

- Is it possible to skip some operations of data processing if their efficiency is not high and, consequently, if it is not worth performing them?

The latter question can be mainly addressed as mentioned below. It is strictly connected with other questions as follows:

- Is it possible to predict the performance of some standard operations of RS data (image) processing?

- What is the accuracy of such a prediction and is this accuracy high enough to undertake a decision to skip carrying out an operation or to set a certain value of some parameter used in the image-processing chain [9]?

This chapter will focus on two typical operations of multichannel RS data processing, namely, filtering and lossy compression. While considering them, the fact that the acquired images are noisy is taken into account. One can argue that noise is not seen in many RS images (or components of these images). This is true, and noise cannot be observed in approximately 80% of the visualized sub-band images of hyperspectral data. This is explained by the peculiarities of human vision, which does not see noise if peak signal-to-noise ratio (PSNR) in a given single-channel (component) image exceeds 32–38 dB. However, recent studies [7], [10-12] have demonstrated that noise is present in all sub-band images and this is due to the principle of operation of hyperspectral imagers.

Moreover, it has been shown in [10], [11] that noise is (can be) of quite a complex nature and the noise acquired in multichannel RS images has specific properties. First, it is signal-dependent [10], [11], [13]. Second, it is of essentially a different intensity (see Abramov et al., 2015 in [14]). More precisely, the wide variation of dynamic range and noise intensity in sub-band images jointly leads to wide limits of signal-to-noise ratio (SNR) in components of multichannel images. This has led to the use of the term "junk bands" [15] and different strategies of coping with noisy channels in multichannel data. Some researchers prefer to use these sub-bands in further processing while others propose to remove them; it is also discussed whether they can be filtered or not [15]. It has been shown that if filtering of these junk bands is efficient, this can improve the classification of hyperspectral data [16]. However, the aforementioned questions concern the efficiency of image preprocessing and its prediction.

The questions raised can be partly answered with the results obtained in recent research. The objective is to show that important performance parameters of image denoising and/or lossy compression can be quickly and quite accurately predicted using simple input parameter(s) and dependences obtained in advance. The obtained results are divided into two parts. The first part deals with the prediction of filtering efficiency. This research has started in 2013 [17] and has its history in a study conducted in [18]. The second part relates to the compression of noisy images [19], [20]. In fact, the results obtained for predicting the parameters of lossy compression can be treated as based on the same principle as that for image filtering and for further research.

Before taking the image performance criteria and preprocessing techniques into consideration, it is important to note the following: first, there are two hypotheses. It is supposed that noise type is known or determined in advance. It is also assumed that its parameters are either known or accurately pre-estimated. It is to be noted that, currently, there are quite a few efficient methods for estimating the parameters of pure additive noise [8], [21-25], speckle noise [26], and different types of signal-dependent noise [10-12], [27], [28]. The noise parameters are taken into account by the most modern filtering techniques that belong to the families of orthogonal-transform-based filters [29-33] and nonlocal filters, for example, block-matching and three-dimensional filtering (BM3D) [34]. The same relates to modern methods of lossy compression of noisy images [19], [35].

Second, we restrict ourselves to consider the image- filtering and compression techniques based on discrete cosine transform (DCT). This is explained using several reasons. DCT is a powerful orthogonal transform widely exploited in image processing. Filters and compression techniques based on DCT are currently among the best [34]. They can be quite easily adapted to the signal-dependent noise directly [32], [36] or equipped with proper variance-stabilizing transformations (VST) [19], [32], [37]. This restriction does not mean that the approach to prediction cannot be applied to other filtering and lossy compression techniques. This approach should be applicable (with certain modifications) but is yet to be thoroughly checked.

Third, in the analysis of the prediction approach, traditional quality metrics are employed such as mean square error (MSE) and peak signal-to-noise ratio (PSNR), as well as some visual quality metrics such as PSNR human visual system masking metric (PSNR-HVS-M) [38]. Behavior and properties of traditional metrics are understood well by those dealing with image processing. Although PSNR-HVS-M is less popular, this is one of the best metrics that takes into account the peculiarities of human visual system (HVS) and that can be calculated for either one component of a multichannel image or a group of components of a multichannel image. It is expressed in dB, and it is usually either slightly smaller than PSNR (for annoying types of distortions like spatially correlated noise) or larger than PSNR (if distortions are masked by texture). This is important since we assume that the processing of multichannel images is carried out either component-wise or in groups of channel images, where a group includes the entire image in marginal case.

Fourth, other criteria of image-processing efficiency, such as classification accuracy, object detectability, etc., are important for the preprocessed RS data. We are unable to predict them, but recent research shows [39] that these criteria are connected with the traditional criteria of image processing. Thus, it is expected that if good values of conventional and HVS metrics are provided due to preprocessing, appropriate classification accuracy and other criteria will be attained.

2. The considered image-performance criteria and preprocessing techniques

This chapter considers the following model of an observed multichannel image:

$$I_{kij}^{\text{noisy}} = I_{kij}^{\text{true}} + n_{kij}(I_{kij}^{\text{true}}), i = 1,...,I, j = 1,...,J, k = 1,...,K \tag{1}$$

where I_{kij}^{noisy} is ij-th sample of noisy (original) k-th component of a multichannel image, n_{kij} denotes the ij-th value of the noise in k-th component statistic, which is, in general, supposed to be dependent on the true image value I_{kij}^{true} in this voxel (3D pixel), I and J define the image size, and K denotes the number of channels. It is also assumed that the images $\{I_{kij}^{\text{true}}\}$ and $\{I_{k+1\,ij}^{\text{true}}\}$ are strongly correlated and they have similar dynamic ranges D_k and D_{k+1} determined as $D_k = I_k^{\text{max}} - I_k^{\text{min}}$, where I_k^{max} and I_k^{min} are maximal and minimal values in k-th channel image, respectively. It is also possible to assume that noise is of the same type and neighbor channels have quite close values of input MSEs (equal to noise variance σ_k^2 if the noise is pure additive) as follows:

$$MSE_k^{\text{inp}} = \sum_{i=1}^{I}\sum_{j=1}^{J}(I_{kij}^{\text{noise}} - I_{kij}^{\text{true}})^2 / (IJ), k = 1,...,K \tag{2}$$

and input PSNR

$$PSNR_k^{\text{inp}} = 10\log_{10}(D_k^2 / MSE_k^{\text{inp}}), k = 1,...,K. \tag{3}$$

The same assumptions are valid for input $PSNR - HVS - M_k^{\text{inp}}$ determined similarly to expression (3) with the difference that MSE_k^{inp} is replaced by $MSE_{HVS\,k}^{\text{inp}}$, which is a special kind of weighted MSE calculated in spectral (DCT) domain considering the masking effects [38]. The aforementioned assumptions are valid for color red, green, blue (RGB) images [27], multispectral and hyperspectral RS images [14], [40], dual polarization, and multifrequency radar images [2]. These properties can be effectively exploited in multichannel image preprocessing [39].

After applying a considered filter, one obtains a filtered image $\{I_{k\,ij}^{f}\}$, $i=1, ..., I$, $j=1, ..., J$, $k=1, ..., K$ that is supposed to be closer to $\{I_{k\,ij}^{true}\}$, $i=1, ..., I$, $j=1, ..., J$, $k=1, ..., K$ according to a chosen metric (a quantitative criterion). These output metrics are calculated as

$$MSE_k^{out} = \sum_{i=1}^{I}\sum_{j=1}^{J}(I_{k\,ij}^{f} - I_{kij}^{true})^2 / (IJ), k=1,...,K, \tag{4}$$

$$PSNR_k^{inp} = 10\log_{10}(D_k^2 / MSE_k^{out}), k=1,...,K. \tag{5}$$

Output $PSNR - HVS - M_k^{out}$ is determined similarly to (5).

Then, one has to characterize the efficiency of filtering. One way to do this is to use

$$\kappa = MSE_k^{out} / MSE_k^{inp}, \tag{6}$$

$$IPSNR_k = PSNR_k^{out} - PSNR_k^{inp}, \tag{7}$$

$$IPHVSM_k = PSNR\text{-}HVS\text{-}M_k^{out} - PSNR\text{-}HVS\text{-}M_k^{inp}. \tag{8}$$

Small values of the ratio in expression (6) and large values of expressions (7) and (8), both expressed in dB, are evidence in favor of efficient filtering.

Similarly, after lossy compression, one obtains $\{I_{k\,ij}^{c}\}$, $i=1, ..., I$, $j=1, ..., J$, $k=1, ..., K$. It is usually supposed that for a larger compression ratio (CR), the quality of compressed image is worse. This is true for lossy compression of noise-free images where more distortions are introduced for a larger CR. However, in lossy compression of noisy images, the situation is specific [41]. Lossy compression results in certain filtering (noise removal) effect under certain conditions. Due to this filtering effect, it is possible that

$$MSE_k^{c} = \sum_{i=1}^{I}\sum_{j=1}^{J}(I_{k\,ij}^{c} - I_{kij}^{true})^2 / (IJ), k=1,...,K \tag{9}$$

occurs to be less than MSE_k^{inp}. Then, the compression method parameter (quantization step (QS), scaling factor (SF) or bits per pixel (bpp) depending upon a coder used) for which MSE_k^c falls into global minimum is called optimal operation point (OOP). This parameter is important and needs additional explanation. Fig. 1(a) presents the dependences of

$$PSNR_k^c = 10\log_{10}(D_k^2 / MSE_k^c), k = 1,...,K \qquad (10)$$

on QS for the lossy DCT-based coder AGU [42] for two known gray-scale test images Airfield (Fig. 1(b)) and Frisco (Fig. 1(c)) corrupted by additive white Gaussian noise (AWGN) with variance $\sigma^2 = 100$. The test image Frisco has a simpler structure – it contains more homogeneous image regions that correspond to sea surface. Due to this, the filtering effect of lossy compression is larger and the dependence has an obvious global maximum (i.e., the OOP), according to $PSNR^c$, since maximum of $PSNR^c$ corresponds to minimum of MSE^c. Formally, there is no OOP for the other test image Airfield, but the dependence $PSNR^c(QS)$ has local maximum. Both aforementioned maxima take place for $QS_{OOP} \approx 4\sigma$, which is a recommended choice for the coder AGU [43].

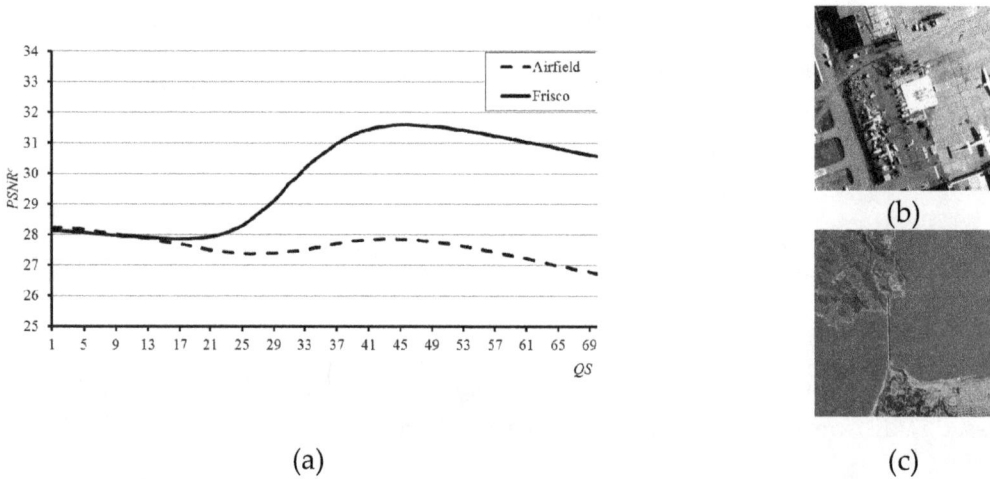

Figure 1. Dependences $PSNR^c(QS)$ for the coder AGU (a) and test images Airfield (b) and Frisco (c) corrupted by AWGN with noise variance equal to 100

The lossy compression in the neighborhood of OOP has obvious advantages. Compressed images have high quality, and, at the same time, they have CR considerably larger than for lossless compression [9], [44]. Because of these benefits, the lossy compression of noisy images in the OOP neighborhood is considered. If OOP does not exist, nevertheless, the recommended setting $QS_{OOP} \approx 4\sigma$ can be considered. If noise is signal dependent and VST is not used, the setting is $QS_{OOP} \approx 4\sigma_{equiv}$ where $\sigma_{equiv}^2 = MSE^{inp}$. Then, in OOP, one has parameters

MSE_k^{OOP}, $PSNR_k^{OOP}$, $PSNR\text{-}HVS\text{-}M_k^{OOP}$ and it is possible to determine for them the following metrics (parameters characterizing compression performance):

$$\kappa = MSE_k^{OOP} / MSE_k^{inp}, \tag{11}$$

$$IPSNR_k = PSNR_k^{OOP} - PSNR_h^{inp}, \tag{12}$$

$$IPHVSM_k = PSNR\text{-}HVS\text{-}M_k^{OOP} - PSNR\text{-}HVS\text{-}M_k^{inp}, \tag{13}$$

where $MSE_k^{OOP} / MSE_k^{inp} < 1$ and positive $IPSNR_k$ or $IPHVSM_k$ mean that OOP exists according to the corresponding metric.

Certainly, there are also other valuable performance criteria. For image pre-filtering, it is important to know the computational efficiency of the denoising method and how easily it can be implemented, especially onboard. For image lossy compression, it is important to know CR provided and how easily it can be attained. To partly address these issues, the filtering and compression techniques are briefly described.

DCT-based filtering [18], [30] is performed in a block-wise manner, where 8 × 8 pixels are a typically set block size. Filtering can be performed with nonoverlapping, partly overlapping, and fully overlapping blocks. In the latter case, filtering efficiency (expressed in improvement of PSNR ($IPSNR$) or improvement of PSNR-HVS-M ($IPHVSM$)) is the highest but more computations are needed. Nevertheless, the filter is very fast since it is possible to use fast algorithms and to parallelize computations.

There are three main steps in processing: direct 2D DCT in each block; thresholding of DCT coefficients; inverse DCT applied to thresholded DCT coefficients; then, the filtered data from overlapping blocks are aggregated. Within this structure, different variants of thresholding are possible but employing hard thresholding is preferred, where DCT coefficient values remain unchanged if their amplitudes exceed a threshold or are assigned zero values otherwise. If one deals with AWGN, the threshold is set fixed as

$$T = \beta\sigma. \tag{14}$$

For spatially uncorrelated signal-dependent noise with *a priori* known or accurately pre-estimated dependence of local standard deviation on local (block) mean $\sigma_{loc} = f(\bar{I}_{bl})$, one has to set a locally adaptive threshold:

$$T_{bl} = \beta f(\bar{I}_{bl}). \tag{15}$$

Finally, for spatially correlated and signal-dependent noise with *a priori* known or properly pre-estimated normalized DCT spectrum W_{qs}^{norm}, $q=0, ..., 7$, $s=0, ..., 7$, where qs are indices of DCT coefficients in blocks [33], the thresholds are locally adaptive and frequency dependent:

$$T_{bl}(q,s) = \beta f(\overline{I}_{bl})\sqrt{W_{qs}^{norm}}. \tag{16}$$

In expressions (14–16), β is the parameter. Depending upon the image complexity and noise intensity, its optimal value can vary a little [18], but the recommended choices are $\beta = 2.6$ to provide good filtering according to *IPSNR* and $\beta = 2.3$ to ensure quasi-optimal denoising according to *IPHVSM*. In further studies, $\beta = 2.6$ will be used. A 3D version of the DCT–based filter [39] performs similarly. The difference is that the blocks are 3D, of size $8 \times 8 \times K_{gr}$, where $K_{gr} \leq K$ denotes a channel group size.

Conventional BM3D [34] is a more sophisticated denoising method. It presumes search for similar patches (blocks), with their joint processing in a 3D manner using DCT and Haar transform, and post-processing stage. This filtering principle, originally designed to cope with AWGN in gray-scale images, has been later adapted to the cases of signal-dependent noise after a proper VST [37], spatially correlated noise [45] and color (three-channel) images corrupted by AWGN [46]. The BM3D and its modifications provide a slightly better performance than the corresponding modifications of the conventional DCT-based denoising by the expense of considerably more extensive computations.

The lossy compression technique called AGU [42] is based on DCT in 32×32 pixel blocks, a more efficient (compared to JPEG) coding of quantized DCT coefficients and post-processing to remove the blocking artifacts after decompression. This coder is quite simple but slightly more efficient than JPEG 2000) or set partitioning in hierarchical trees (SPIHT) in rate/distortion sense. This coder has 3D version [19] and CR for both 2D and 3D versions is controlled (changed) by QS.

3. Prediction of filtering efficiency

The main idea of filtering efficiency prediction is the following [17]. Suppose there is some input parameter(s) able to jointly characterize image complexity and noise intensity and also there is some output parameter(s) capable of adequately describing the image denoising efficiency. Assume that there is a rather strict connection between these input and output parameters that allows predicting output value(s) having input value(s).

An additional assumption (and requirement to prediction) is that input parameter(s) have to be calculated easily and quickly enough, faster than denoising itself (otherwise, the prediction becomes useless). If all these assumptions are valid, it becomes possible to determine a predicted output value before starting image filtering and to decide whether it is worth filtering a given image (component) or not. Another decision can relate to setting parameter(s) of a used filter. For example, if a processed image seems to be textural (having high complexity), parameter(s) of a used filter can be adjusted to provide better edge/detail/texture preservation. For example, the parameter β for the DCT-based filter can be set equal to 2.3.

Keeping these general principles in mind, we have to address several tasks:

- What is a good (in the best case, optimal) input parameter (or a set of parameters)?

- What is a good (proper, acceptable) output parameter (or a set of parameters) that allows to characterize the filtering efficiency adequately and to undertake a decision (on using filtering or not, on setting a filter parameter, etc.)?

- How to get dependence between output and input parameters and how accurate it is?

These questions are partly answered below and the outcomes obtained in design and performance analysis of prediction techniques are described. We believe that a partial answer to the second question is the following. The ratio in expression (6) as well as the parameters $IPSNR_k$ and $IPHVSM_k$ (especially if analyzed jointly) are able to provide the initial insights (characterization) of filtering efficiency. Note that expressions (6) and (7) are mutually dependent metrics and $IPSNR_k = 10\log_{10}(MSE_k^{inp}/MSE_k^{out})$. Thus, they can be used as output parameter(s) at the current stage of research.

3.1. Input and output parameter sets testing and comparison

Based on the outcomes of the study [18], Abramov et al. in 2013 [17] observed that there is dependence between efficiency of filtering expressed by (6) and simple statistics of DCT-coefficients determined in 8×8 blocks. Two probability parameters have been considered. The first one denoted as $P_{2\sigma}$ is the mean probability that the amplitudes of DCT coefficients are not larger than 2σ, where σ denotes the standard deviation of additive white Gaussian noise. This parameter originated from analogies with known sigma filter [47]. The second parameter denoted as $P_{2.7\sigma}$ is the mean probability that the amplitudes of DCT coefficients are larger than 2.7σ. Here, there is an obvious analogy with hard thresholding in DCT-based filter, where the recommended $\beta = 2.7$. At the starting point, Abramov et al., 2013 had no idea on the optimality of input parameters. The objective was just to check whether the prediction is possible, in principle, using a restricted set of test gray-scale images (18) and standard deviations of AWGN (5, 10, 15). The data have been presented as scatterplots, where the Y-axis reflects the ratio in expression (6) and X-axis corresponds to a considered statistical (input) parameter (either $P_{2\sigma}$ or $P_{2.7\sigma}$). These scatterplots are represented in Fig. 2. Obviously, the scatterplots' points are clustered well along the fitted lines (for easy fitting, second-order polynomials were used). Interestingly, small $P_{2\sigma}$ and large $P_{2.7\sigma}$ correspond to complex structure images corrupted by low-intensity noise. In this case, efficiency of image filtering is low (the ratio in expression (6) is close to unity, see Fig. 2). Note that this is in agreement with the theory of filtering [48], [49]. It shows that efficient filtering of textural images is problematic for any existing filters including the most sophisticated nonlocal ones [34].

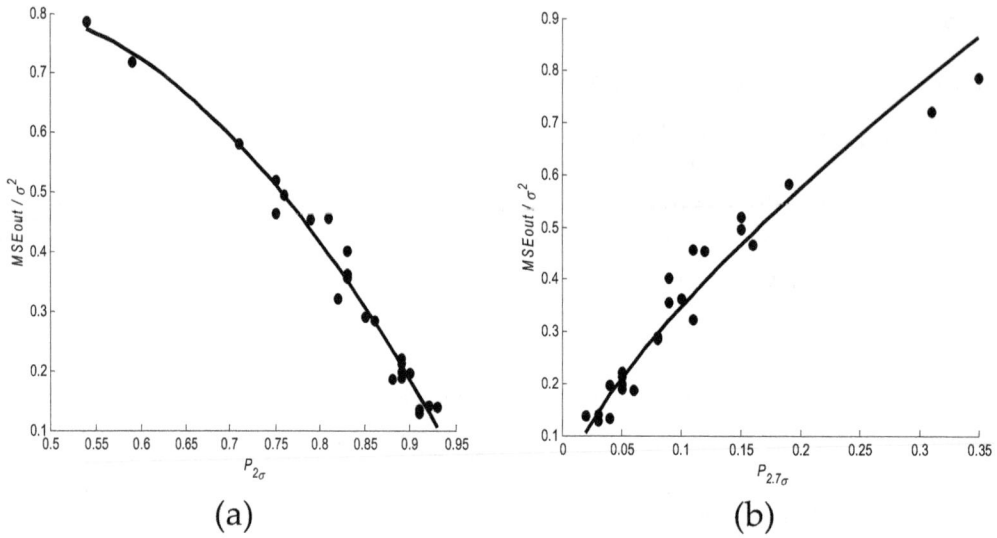

Figure 2. Examples of scatterplots and curve fitting into them for $P_{2\sigma}$ (a) and $P_{2.7\sigma}$ (b)

The results of the study conducted in [17] have also shown the following. First, quality of fitting has to be characterized quantitatively. For this purpose, the approach [50] works well. It provides the parameter (coefficient of determination) R^2 that tends to unity for perfectly fitted curves and root mean square error (RMSE) of fitting that should be as small as possible. These parameters are strictly connected with prediction accuracy. For perfectly determined $P_{2\sigma}$ or $P_{2.7\sigma}$, RMSE of fitting directly describes the accuracy of prediction.

The conclusions drawn in [17] can be recalled here. First, the prediction of filtering efficiency for BM3D is less accurate than for the conventional DCT-based filter. This conclusion has been confirmed in later studies. This is associated with the use of two denoising mechanisms (DCT denoising and similar block search with their joint processing), where the latter mechanism has no connection to DCT statistics. Second, although the prediction accuracy for both $P_{2\sigma}$ and $P_{2.7\sigma}$ is quite good ($R^2 > 0.9$ and RMSE < 1.0), the probability $P_{2\sigma}$ provides sufficiently better prediction (quality of fitting) than $P_{2.7\sigma}$. This shows that the use of other input parameters is possible. Third, different types of fitting functions (polynomials, power and exponential functions) were able to provide approximately the same quality of fitting (for example, the fitted curve in Fig. 2(a) is $\kappa = -2.63P_{2\sigma}^2 + 2.15P_{2\sigma} + 0.38$, for the BM3D filter, the obtained function of $P_{2.7\sigma}$ is $\kappa = 1.86P_{2.7\sigma}^{0.73}$). Thus, certain reserves in improving the fitting accuracy "are hidden" in choosing an approximating curve and its parameters. Fourth, it has also been shown that the probabilities $P_{2\sigma}$ and $P_{2.7\sigma}$ can be determined with appropriate accuracy from analysis of not all possible overlapping blocks but from partly or even nonoverlapping blocks if their total number is not less than 300...500. This additionally accelerates the prediction compared even to conventional DCT-based filtering.

There are also observations understood later (in two recent years). First, there should be some restrictions imposed on the approximating function. For example, it is clear that the ratio in expression (6) cannot be negative. It is also clear that an approximating (fitting) function should be determined for all possible values of its arguments. Since the probabilities serve as arguments, they can vary from zero to unity. Meanwhile, arguments in both scatterplots in Fig. 2 vary in narrower limits. Besides, it could be good for curve fitting to have point arguments with approximately uniform density.

These requirements have been satisfied by using considerably more test images (including highly textural ones) and a wider set of noise standard deviations (including quite small ones). This has allowed obtaining scatterplot points for small $P_{2\sigma}$ and large $P_{2.7\sigma}$.

Examples of the obtained scatterplots and fitted curves for the DCT-based denoising are shown in Fig. 3. As it is seen, fitting is rather good and coefficient of determination is approximately 0.95 (see the details below). We believe these are already good results that allow practical recommendations. For example, it is clearly seen that there is no reason to carry out filtering if $P_{2\sigma}$ is smaller than 0.5 since the benefit obtained due to denoising is negligible (approximately 1 dB or less). Prediction itself is carried out as follows. Having the fitted curves obtained in advance as described above, it is needed to calculate $P_{2\sigma}$ or $P_{2.7\sigma}$ for a given image before filtering and to substitute it as argument into the approximating function to calculate a desired metric that characterizes the predicted denoising efficiency.

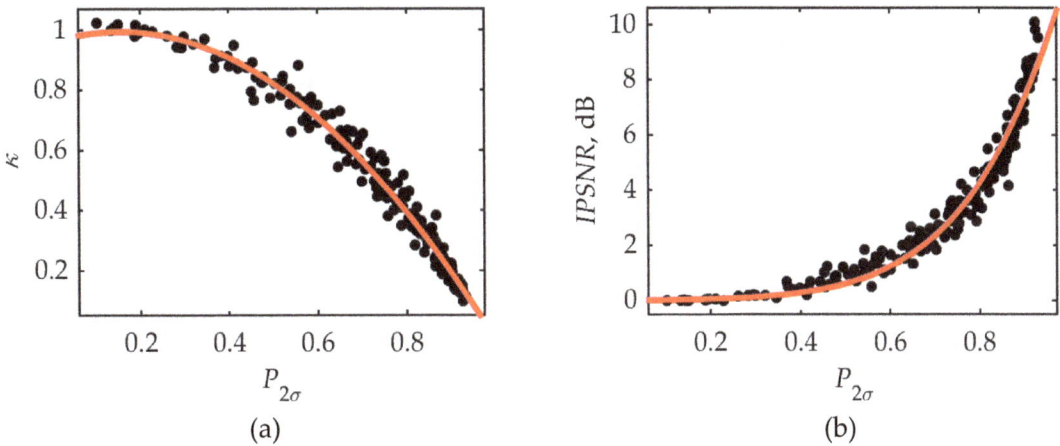

(a)   (b)

Figure 3. Scatterplots of κ (a) and IPSNR (b) on $P_{2\sigma}$ and the fitted curves

Expressions for the obtained approximations for the DCT filter are as follows (we give only the functions of $P_{2\sigma}$, more details can be found in [51]):

$$\kappa = -1.45P_{2\sigma}^2 + 0.45P_{2\sigma} + 0.96, \tag{17}$$

$$IPSNR = 100 * \exp\left(-\left(\frac{P_{2\sigma} - 1.92}{0.63}\right)^2\right), \tag{18}$$

$$IPHVSM = 100 * \exp\left(-\left(\frac{P_{2\sigma} - 2.08}{0.67}\right)^2\right). \tag{19}$$

The values of R^2 are presented in Table 1. The analysis confirms that it is better to use $P_{2\sigma}$ than $P_{2.7\sigma}$. Prediction of κ is slightly more accurate than the prediction of $IPSNR$. However, the prediction of $IPHVSM$ is worth improving.

Metric	$P_{2\sigma}$	$P_{2.7\sigma}$
K	0.978	0.955
$IPSNR$	0.963	0.935
$IPHVSM$	0.82	0.78

Table 1. Goodness of fit (R^2) of the obtained approximations

It has been discovered that not only the mean of local (block) estimates of probability $P_{2\sigma}$ is connected with predicted metrics [51], but the other statistical parameters of the distribution of local estimates can also be exploited to improve prediction. The general framework to obtain an estimate of a predicted metric by multiparameter fitting is described by the following formula:

$$Metric_{est} = a * \exp\left(\sum_{i=1}^{n} b_i O_i(P)\right), \tag{20}$$

where a and b_i are approximation factors, O_i. $i = 1,...,n$, is some parameter of distribution, n defines the number of such parameters. As O_i, it is possible to use the distribution mean, median, mode, variance, skewness, and kurtosis. The factors a and b_i, $i = 1,...,n$ have to be obtained in advance by multidimensional (n-dimensional) regression.

The results of using multidimensional regression are presented in Table 2. The abbreviations used are the following: M – mean; Var – variance; Med – median, Mod – mode; K – kurtosis; S – skewness; all calculated for a set of local estimates of probability $P_{2\sigma}$. The results are given for both considered filters for the metrics $IPSNR$ and $IPHVSM$. Only the best sets for n from 1 to 5 are presented since the joint use of all considered parameters is less efficient than five input parameters employed together.

Filter	Metric	Statistical Parameters	R^2
DCT filter	IPSNR	M	0.963
		M, Var	0.971
		M, Var, Mod	0.974
		M, Var, Mod, K	0.976
		M, Var, Med, Mod, S	0.977
	IPHVSM	Med	0.848
		M, Var	0.923
		M, Var, Med	0.926
		M, Var, Med, S	0.927
		M, Var, Med, Mod, S	0.928
BM3D	IPSNR	M	0.95
		M, Var	0.955
		M, Var, Mod	0.959
		M, Var, Mod, S	0.961
		M, Var, Med, Mod, S	0.961
	IPHVSM	Med	0.845
		M, Var	0.905
		M, Var, S	0.905
		M, Var, S, K	0.909
		M, Var, Med, S, K	0.917

Table 2. Goodness of the best multiparameter fit for $P_{2\sigma}$

The conclusions are the following. The use of more input parameters leads to larger (better) R^2 for both filters and both metrics. The benefit of using several input parameters instead of one is quite small for *IPSNR*, where R^2 for one-parameter prediction is already quite high. Meanwhile, for the visual quality metric *IPHVSM*, the improvement is quite large. Interestingly, the use of median of local estimates instead of the mean considerably improves prediction (compare the data in Tables 2 and 1) for *IPHVSM* for the DCT-based filter and $P_{2\sigma}$.

Filter	Metric	a	b_1	b_2
DCT filter	IPSNR	0.023	6.338	7.459
	IPHVSM	$2.225*10^{-4}$	10.81	37.14
BM3D	IPSNR	0.019	6.591	6.849
	IPHVSM	$5.324*10^{-5}$	12.42	41.36

Table 3. Coefficient values of the obtained approximations for $P_{2\sigma}$

More input parameters provide better prediction. At the same time, more time is needed for calculation of input parameters (although their calculation is not difficult). Then, a compromise solution could be the use of the dependence of the type

$$\text{Metric}_{est} = a^* \exp\left(b_1 \text{ mean}(\hat{P}_{2\sigma \text{ loc}}) + b_2 \text{ var}(\hat{P}_{2\sigma \text{ loc}})\right), \tag{21}$$

where $\hat{P}_{2\sigma \text{ loc}}$ denotes the local estimates of probabilities obtained in blocks. The approximation coefficients for all cases are presented in Table 3.

The expression (20) is not the only way to combine several input parameters into a joint output. Neural networks (NN) are known to perform this task rather well and to be good approximators [52]. This property has been used by us in [53] to make the neural network predict the considered metrics based on multiple input parameters. The obtained results are practically the same as in Table 3. Therefore, there is no need to use a more complex NN approximator instead of expression (20).

A more reasonable solution is to look for better input parameters. Such a study has been conducted in [51]. It has been shown that the probability $P_{0.5\sigma}$ is more informative than $P_{2\sigma}$, that $P_{0.5\sigma}$ is the mean probability where the magnitudes of DCT coefficients in blocks are smaller than 0.5σ. Theoretically, for Gaussian distribution, this probability does not exceed 0.38. Gaussian distribution takes place for DCT coefficients of AWGN. Thus, the mean $P_{0.5\sigma}$ approaches to 0.38 only if a considered image is "very homogeneous" and noise is intensive. This is postulated in further studies.

The obtained results for multiparameter fitting are presented in Table 4. The abbreviations are the same as in Table 2. The first observation is that even for one parameter (mean of local probabilities), the values R^2 are sufficiently better than the corresponding values for $P_{2\sigma}$. Again the results for the BM3D filter are slightly worse than for the DCT-based filter and the results of predicting IPHVSM are worse than for predicting IPSNR. Again the use of only two input parameters, mean and variance of local estimates, seems to be a good practical choice. Thus, the best parameters of the function (21) are presented for this case in Table 5. Besides, we give an example of scatterplot fitting by 2D surface (function) for two-parameter case of using mean and variance of local estimates of the considered probability for predicting IPHVSM (see Fig. 4).

Filter	Metric	Statistical Parameters	R^2
DCT filter	IPSNR	M	0.986
		M, Var	0.989
		M, S, K	0.989
		M, Med, S, K	0.989
		M, Var, Med, Mod, S	0.99
	IPHVSM	Mod	0.844
		M, Var	0.944
		M, Var, Mod	0.949
		M, Var, Mod, S	0.951
		M, Var, Med, Mod, S	0.952
BM3D	IPSNR	M	0.975
		M, Var	0.977
		M, Var, S	0.978
		M, Var, Med, S	0.978
		M, Var, Med, Mod, S	0.978
	IPHVSM	Mod	0.852
		M, Var	0.935
		M, Var, Mod	0.939
		M, Var, Mod, S	0.941
		M, Var, Med, Mod, S	0.941

Table 4. Goodness of the best multiparameter fit for $P_{0.5\sigma}$

Filter	Metric	a	b_1	b_2
DCT filter	IPSNR	0.168	10.8	19.28
	IPHVSM	0.01	15.66	144.3
BM3D	IPSNR	0.148	11.33	17.7
	IPHVSM	0.004	18.25	161.7

Table 5. Approximation coefficients values of obtained approximations for $P_{0.5\sigma}$

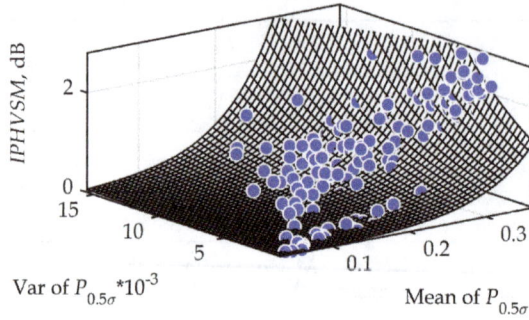

Figure 4. Scatterplot of *IPHVSM* for the DCT-based filter efficiency on statistics of $P_{0.5\sigma}$ and the fitted surface

3.2. Analysis for signal-dependent and spatially correlated types of noise

Let us define the models of signal-dependent noise used. According to a first model [7], [11], the expression (1) transforms to

$$I_{kij}^{noise} = I_{kij}^{true} + N_{kij}^{SI} + N_{kij}^{SD}, \tag{22}$$

where N_{kij}^{SI}, N_{kij}^{SD} denote signal-independent (SI) and signal-dependent (SD) noise components. Both the noise components in expression (22) are assumed zero mean, spatially uncorrelated and Gaussian. Then, the model for the noise variance is $\sigma_{kij}^2 = \sigma_k^2 + \gamma I_{kij}^{true}$, where σ_k^2 is the SI noise variance and γ is the SD noise parameter (which is usually between zero and unity). A second model [2] presumes purely multiplicative noise with $I_{kij}^{noise} = I_{kij}^{true}\mu_{kij}$, where μ_{kij} denotes unity mean random factor with variance $\sigma_{\mu k}^2$ that is within the limits from 0 to 1. It is supposed for both the models that the noise is spatially uncorrelated.

As mentioned in Section 2 (expression no. 15), the local threshold is set as $T_{bl} = \beta\sqrt{\sigma_0^2 + \gamma\bar{I}_{bl}}$ for signal-dependent noise (expression no. 22) and as $T_{bl} = \beta\sigma_\mu\bar{I}_{bl}$ for pure multiplicative noise. In addition to modifying the filtering algorithm, we need to modify the algorithm of input parameter calculation. Then, the local probability estimate has to consider the local variation of noise standard deviation. For instance, the local estimate of probability $P_{2\sigma}$ is obtained as

$$\hat{P}_{2\sigma}^{bl} = \sum_{q=0}^{7}\sum_{s=0}^{7}\delta_{qs} / 63, \tag{23}$$

where $\delta_{qs} = 1$, if $|D_{qs}| \leq 2\sigma_{bl}$ and 0 otherwise (σ_{bl} is equal to $\sqrt{\sigma_0^2 + \gamma\bar{I}_{bl}}$ or to $\sigma_\mu\bar{I}_{bl}$ depending upon a model used). DC component of DCT coefficients in blocks is not taken into account as it always exceeds the local threshold.

Some of the results of studies in our papers [54], [55] are presented next. One aspect that was specially addressed in these studies was to check the influence of an image set used in forming a scatterplot. In fact, two scatterplots have been formed separately: for the set of standard images used in optical image processing as Baboon, Barbara, Lena, etc., and for the set of images called "Remote Sensing" as Frisco, Diego, etc. The reason for such study was the following fact. Some people from RS community are categorically against using standard gray-scale test images in their studies although there are no commonly accepted sets of test RS images.

The methodology of obtaining scatterplot was modified a little. For the noise expression model (22), three different cases were modeled: prevailing influence of SI noise, dominant influence of SD noise, and comparable contribution of both components. As a result, a wide range of mean $P_{2\sigma}$ has been provided. Scatterplot points that belong to different image sets are indicated by different signs (and different colors). There are also two fitted curves. We believe there is no essential difference between the scatterplots and fitted curves. Thus, it can be concluded that the prediction is quite universal and suitable for conventional gray-scale optical images and component-wise (single-channel) RS images. Moreover, it has been shown in a study [55] that prediction is valid for single-look SAR images corrupted by fully developed spatially uncorrelated speckle. It is also possible to compare the results in Fig. 5 with the data in Fig. 3(b). They are very similar. Fig. 4 shows that *IPSNR* is approximately 1 dB or less for $P_{2\sigma}$ approximately 0.5 and then denoising is practically useless. Meanwhile, if *IPSNR* is approximately 4 dB for $P_{2\sigma}$ approximately 0.8, then the use of filtering is expedient. The parameter R^2 for both fitting curves in Fig. 5 is approximately 0.96, that is, the prediction is approximately as good as for AWGN case. Again, the results for $P_{2\sigma}$ are better than for $P_{2.7\sigma}$; fitting for *IPSNR* is more accurate than for *IPHVSM*. Improved fitting by means of using multiple input parameters has not been investigated yet.

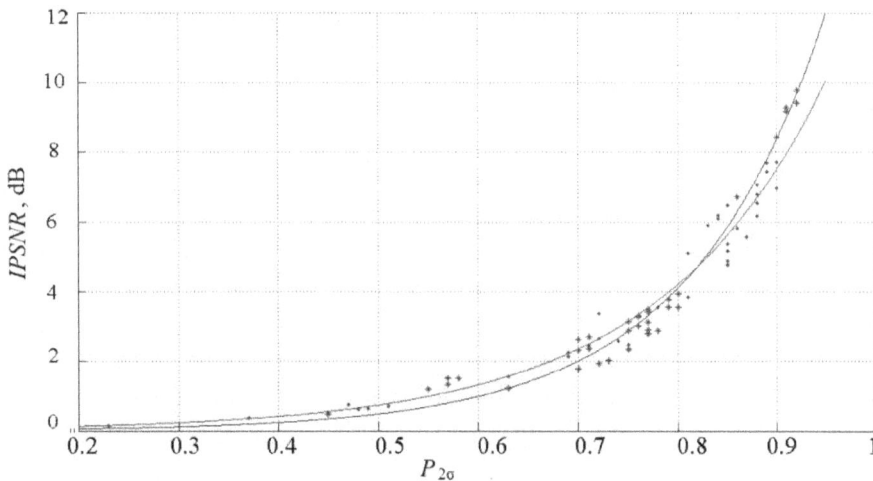

Figure 5. Scatterplots of *IPSNR* for the DCT-based filter efficiency on statistics of $P_{2\sigma}$

Two examples of image processing are presented here. Fig. 6(a) represents the noisy image Frisco, where noise parameters are $\sigma_0^2=100$; $\sigma^2=100$, and $\gamma=0.2$. The output image for the DCT-based filter is presented in Fig. 6(b). The effect of denoising is obvious. Actual provided improvement of PSNR is equal to 9.77 dB. The predicted value for mean $P_{2\sigma}=0.92$ is approximately 9.5 dB (see the blue fitted curve in Fig. 5), that is, there is good agreement of attained and predicted values. Prediction shows that it is worth applying denoising in this case.

For a real-life data, it is impossible to determine true values of the considered metrics characterizing filtering efficiency. However, it is possible to analyze the predicted values and denoising results visually. For fragments of sub-band images of hyperspectral sensor, Hyperion, such analysis was done. For example, noise parameters of the expression model (22) have been blindly estimated [11]. The noisy image for the 13th sub-band of the set EO1H1800252002116110KZ is depicted in Fig. 7(a). Noise is clearly seen. The prediction of *IPSNR* is approximately 8.5 dB and *IPHVSM* is approximately 5.7 dB. Thus, it is expedient to perform denoising. The denoised image is presented in Fig. 7(b). As can be seen, its quality has very much improved due to filtering.

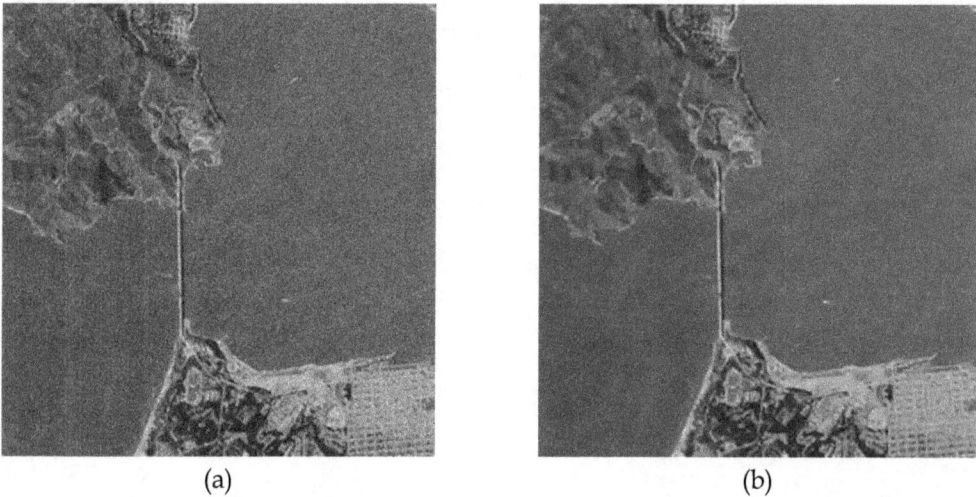

(a)   (b)

Figure 6. Noisy (a) and output (b) images Frisco

The sub-bands 13...22 are considered for two sets of Hyperion data. The values *IPSNR* are always larger than *IPHVSM*. This means that it is harder to provide an improvement of image visual quality than to gain improvement according to standard metrics (MSE, PSNR). For the sub-bands with indices $k = 13...16$, *IPSNR* is always larger than 1.6 dB and *IPHVSM* exceeds 0.6 dB, that is, filtering is desirable. For other sub-bands, as the predicted improvements are small, it is doubtful whether it is worth carrying out filtering. Visual inspection of images in sub-bands with $k = 17...22$ has shown that noise is either hardly noticeable or practically invisible. Positive effect of its removal is partly or fully compensated by edge/detail/texture smearing performed by any filter, even the most sophisticated one [56]. The texture filtering is always problematic and the prediction approach is able to reliably predict this [56].

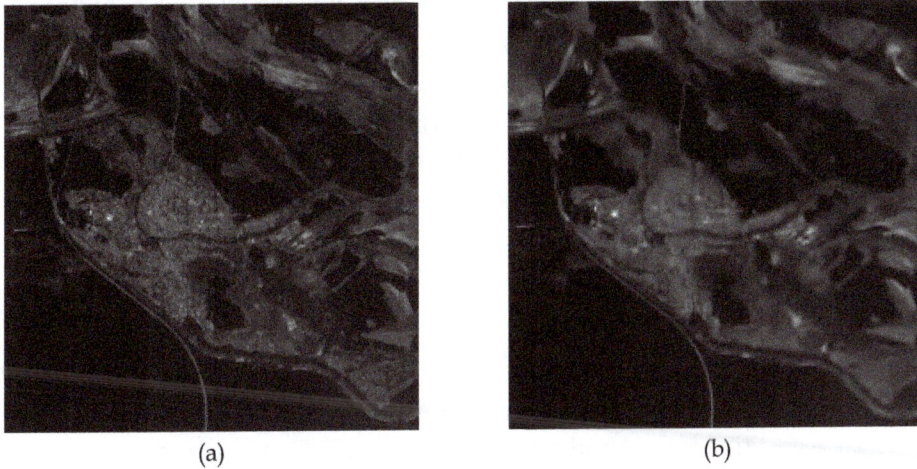

Figure 7. Noisy (a) and output (b) images of 13th sub-band images of Hyperion sensor

Considering certain benefits achieved due to using $P_{0.5\sigma}$ as input parameter, the analysis similar to the one presented in Fig. 5 has been performed. The results are presented in Fig. 8. The noise is signal-dependent and most scatterplot points correspond to the expression model (22). The curve is fitted employing all points (although they relate to optical and RS subsets). Obviously, fitting is very good and, according to quantitative criteria, it is better than for the parameter $P_{2\sigma}$ (Fig. 5). Four black points at the scatterplot in Fig. 8 correspond to one-look SAR images. They fit the curve well and have the arguments close to the maximal potential limit (0.38), where *IPSNR* attains very large values (approximately 10 dB and more).

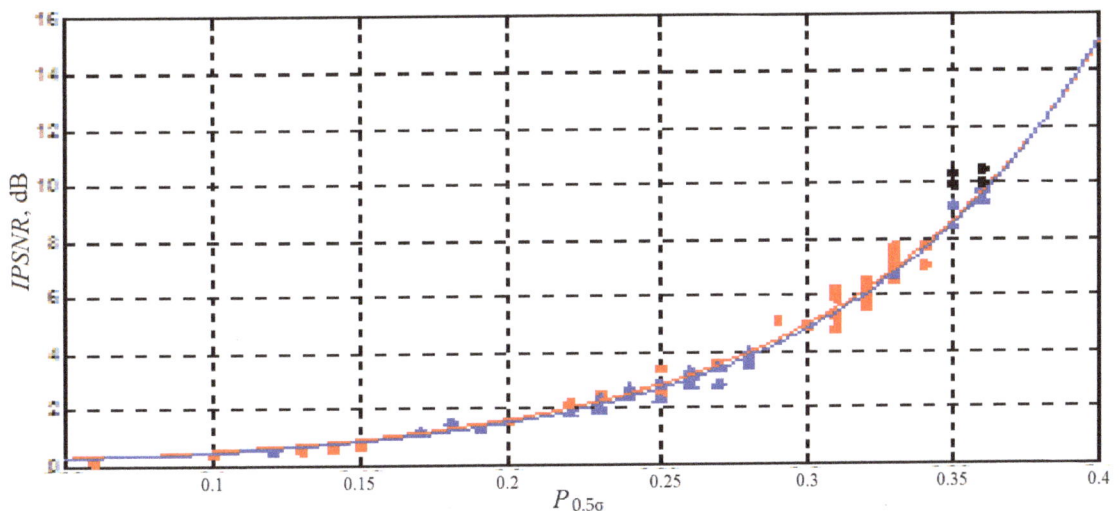

Figure 8. Scatterplots of *IPSNR* for the DCT-based filter efficiency on the statistics of $P_{0.5\sigma}$ (two sets of images and two fitted curves)

Additional studies concentrated on the multi-look SAR images that were corrupted by pure multiplicative noise [57]. Analysis has been done for speckle variance $\sigma_\mu^2 = 0.273/L$, where L denotes the number of looks. Scatterplot points are presented in Fig. 9 for different number of looks. An obvious tendency is that mean $P_{0.5\sigma}$ becomes larger and *IPSNR* increases for smaller number of looks. Other conclusions that can be drawn from analysis in a study in [57] are the following. Prediction is possible for filtering techniques with and without VST, where the prediction quality is better in the latter case. Prediction using different types of functions (polynomial, power, exponential) produce fitting of approximately equal accuracy. Meanwhile, accuracy of prediction is worth improving (RMSE is approximately 1 dB) since it is sufficiently worse than for the case of AWGN.

Understanding that, in practice, noise can be spatially correlated [33], the case of spatially correlated noise – additive in [45] and multiplicative in [57] – are also studied. A difficulty of dealing with spatially correlated noise is that there are numerous shapes (and parameter sets) of 2D auto-correlation function or spatial spectrum of such a noise. Thus, studying a particular case of spatially correlated noise gives only limited information on general dependences. Hence, two models of spatially correlated noise (called middle correlation and strong correlation) have been considered [45]. A peculiarity of prediction is that the local estimate of probability $P_{2\sigma}$ is obtained according to expression (23), where, in the general case, $\delta_{qs} = 1$, *if* $|D_{qs}| \leq 2\sigma_{bl}(W(q, s))^{1/2}$ and 0 otherwise (σ_{bl} is the local standard deviation in a considered block; expressions for its derivation depending upon noise model are given above). If the probability $P_{0.5\sigma}$ is used, the condition is $\delta_{qs} = 1$, *if* $|D_{qs}| \leq 0.5\sigma_{bl}(W(q, s))^{1/2}$ and 0 otherwise.

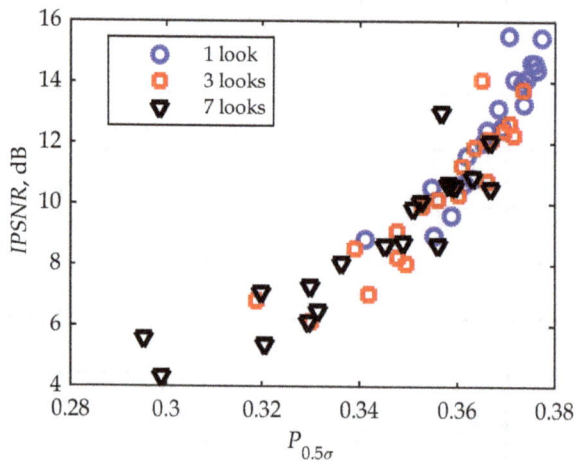

Figure 9. Scatterplot for *IPSNR* vs. mean $P_{0.5\sigma}$ for a part of test images corrupted by spatially uncorrelated speckle

The scatterplots and fitted curves are presented in Fig. 10. The fitted curves are similar and they clearly show that there is no reason to filter images if $P_{0.5\sigma}$ is smaller than 0.15. The difference in the scatterplots for *IPHVSM* and *IPSNR* is that the latter one is more compact and, thus, *IPSNR* can be predicted more accurately. An additional distinctive feature of the plot for

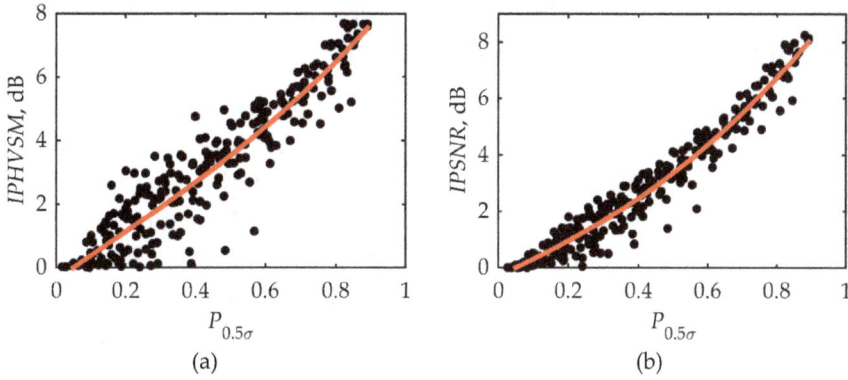

Figure 10. The scatterplots for middle-correlation noise and the fitted curves for *IPHVSM* (a) and *IPSNR* (b)

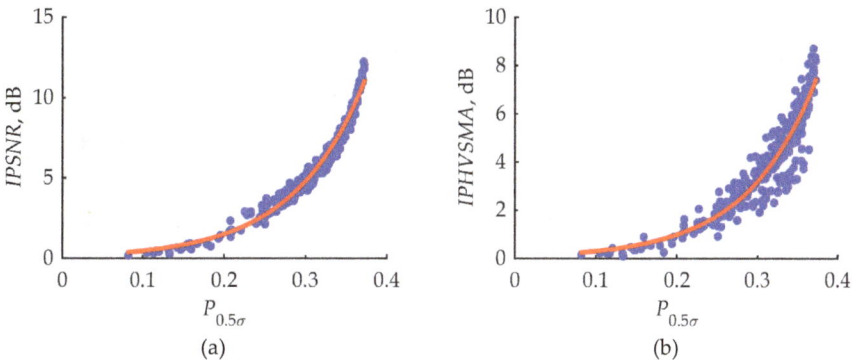

Figure 11. The scatterplots and the fitted curves for *IPSNR* (a) and *IPHVSMA* (b)

IPSNR is that its maximal values are smaller than for AWGN case (data in Fig. 3(b)). The scatterplots for a strong correlation of the noise and the conclusions derived from them are similar.

We have also studied the case of spatially correlated speckle [57]. It has been shown that the prediction seems possible for a spatially correlated noise. However, more research is needed to understand how to select a parameter or several parameters to characterize spatial correlation and how it can be involved in prediction.

Finally, a preliminary research has been carried out for denoising color images corrupted by AWGN with equal variance values in channels [58]. There are two differences in prediction. First, all DCT coefficients in 3D block are subject to analysis for estimating the local probabilities. Second, the metric PSNR-HMA [59], which is a color extension of PSNR-HVS-M, and improvement of this metric due to filtering similar to expression (8) have been used. In addition, instead of BM3D, its color version called C-BM3D has been analyzed [46].

The scatterplots have been obtained and curves were fitted to them (see examples in Fig. 11). As mentioned earlier, filtering is useless for $P_{0.5\sigma} < 0.15$. However, this happens rarely (only for highly textured images when noise standard deviation is small). Another observation is the same as earlier – visual quality can be predicted worse than *IPSNR*. The prediction accuracy for C-BM3D is worse than for 3D DCT filter.

Taking into account our previous experience, the multiparameter input was analyzed with exponential function expressed in (20). Considerable improvement has been reached, especially for *IPHVSMA,* for the 3D DCT filter. For the C-BM3D filter, the positive effect is less. One has R^2 equal to 0.8481 for one input parameter and 0.8555 for four parameters. Again, a reasonable practical solution is to use the mean and variance of local estimates of probability. One more important observation for color image filtering is that $P_{0.5\sigma}$ for 3D filter is larger than for DCT filter applied to components of a processed color image. This again proves that 3D processing of color and multichannel images iiis are potentially more efficient compared to their component-wise denoising.

4. Prediction in lossy compression of noisy images

In this section, the compression of images corrupted by AWGN is considered. Lossy compression is carried out by the aforementioned coder AGU with $QS = 4\sigma$. In this case, OOP may exist or be absent. The task is to predict *IPSNR* and *IPHVSM* and to decide whether OOP exists as well as to predict what CR is.

4.1. Prediction of OOP existence and metrics' values in it

This section shortly describes how the scatterplots were obtained. As in the filtering case, a set of gray-scale test images of different content and complexity was used. AWGN of different intensity has been added and then the obtained images have been compressed by AGU. After this, the parameters (12) and (13) have been calculated as well as $P_{2\sigma}$ for each compressed image. Clearly, all these actions are done off-line before applying the prediction approach in practice.

The obtained scatterplot is presented in Fig. 12. A specific feature of this scatterplot is that it has negative values and they seem to be approximately –3.5 dB for $P_{2\sigma}$ approaching to zero. Therefore, not all fitting functions can be used. The study carried out by Zemliachenko et al. in [44] has shown that the polynomials of the fourth and fifth order usually allow approximating the dependence very well (with R^2 almost equal to unity and RMSE approximately 0.25 for *IPSNR*). As can be seen from the analysis of the scatterplot in Fig. 12, there are quite many images and/or noise variances when OOP does not exist (*IPSNR* is negative). OOP exists with high probability if $P_{2\sigma}$ exceeds 0.82. This can be used as a basis for predicting OOP existence.

The scatterplot for the metric *IPHVSM* is presented in Fig. 13. In some sense, behavior of the fitted polynomial is similar to the one in Fig. 12. There are many values about –4 dB showing that due to lossy compression the visual quality becomes worse. However, this mainly happens for small $P_{2\sigma}$ that corresponds to high complexity images and/or low level of the noise. The visual quality improves for $P_{2\sigma}$ exceeding 0.9 and this takes place for low-complexity images and rather intensive noise.

$$y = 156.3x^5 - 301.3x^4 + 214.3x^3 - 64.64x^2 + 8.65x - 3.44$$

SSE: 5.272
$R - square$: 0.9911
Adjusted $R - square$: 0.9905
RMSE: 0.26

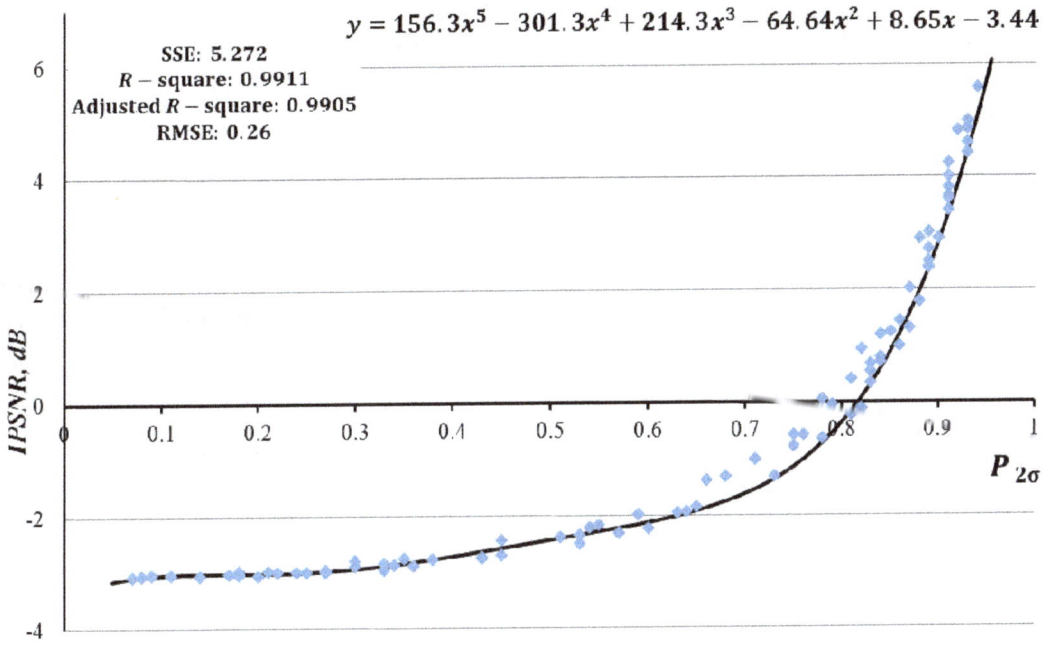

Figure 12. The scatterplot and the fitted curves for $IPSNR$ and the coder AGU

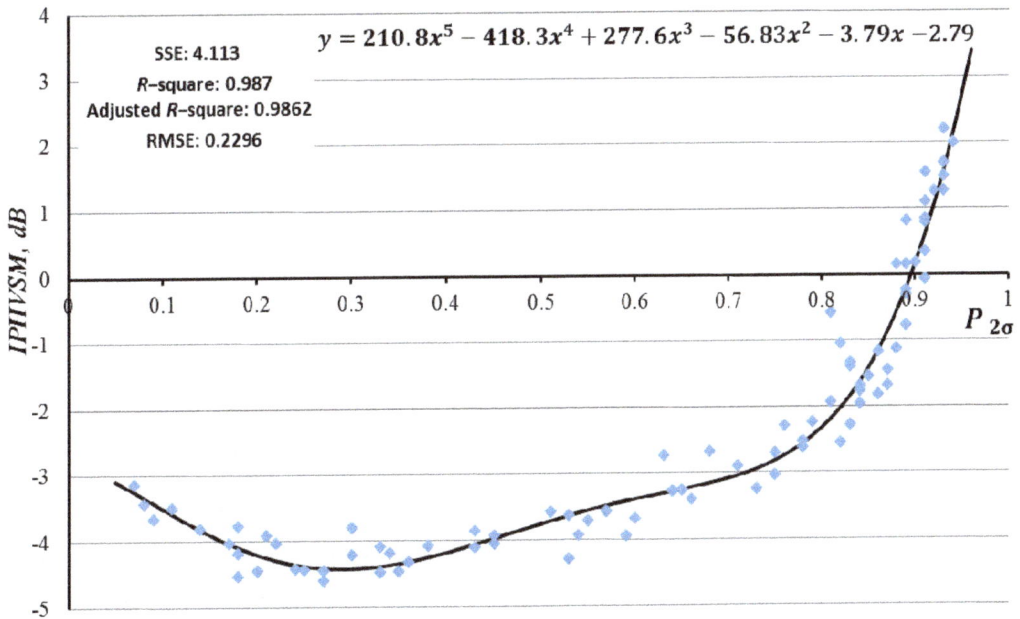

$$y = 210.8x^5 - 418.3x^4 + 277.6x^3 - 56.83x^2 - 3.79x - 2.79$$

SSE: 4.113
R–square: 0.987
Adjusted R–square: 0.9862
RMSE: 0.2296

Figure 13. The scatterplot and the fitted curves for $IPHVSM$ and the coder AGU

Although prediction has been studied by simulations only for images corrupted by AWGN, it can also be applied to images corrupted by a signal-dependent spatially uncorrelated noise under condition that a proper VST is applied to them before compressing. Such VST (a generalized Anscombe transform in this case) provides approximately constant noise variance that usually equals to unity. Thus, QS = 4 is used. This approach has been used for Hyperion data and the results are presented in Fig. 14. There are two groups of sub-bands that are usually not analyzed in Hyperion data since they are too noisy. Thus, the prediction values are not given for all sub-bands. Analysis of the presented values shows that there are only a few sub-bands where it is worth expecting OOP. For most other sub-bands, *IPSNR* is about −3 dB and the ways of dealing with them are considered in a study [44]. One proposition is to set less QS but this leads to smaller CR.

Fig. 15 shows the original and the decompressed images in 110-th sub-band, where decrease of visual quality according to quantitative criteria is predicted. Noise is not seen in the original image and the compression practically does not influence the image quality (in our opinion, both images look the same).

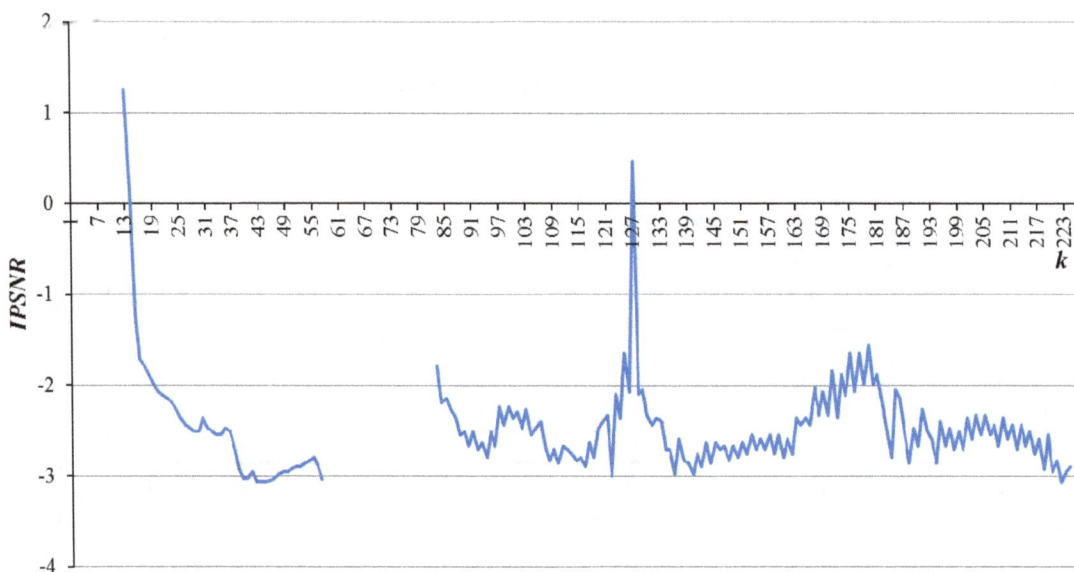

Figure 14. Predicted *IPSNR* for Hyperion hyperspectral image compressed by the coder AGU

A study [44] also presents data for three other DCT-based coders, where two of them are specially suited for providing better visual quality. It is demonstrated that the coder adaptive DCT (ADCT), which exploits the optimized partition schemes [60], provides certain improvements compared to AGU. Meanwhile, DCT coders oriented on improving the visual quality being applied to noisy images do not offer substantial benefits and, moreover, are even less efficient in many practical situations.

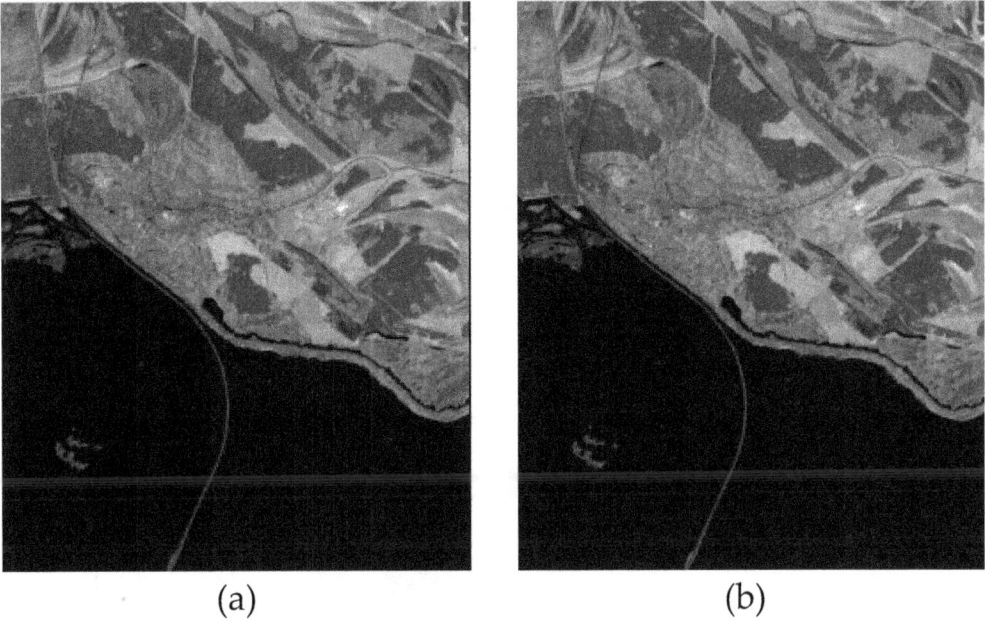

(a)   (b)

Figure 15. The 110-th sub-band images before (a) and after (b) compression

4.2. Prediction of compression ratio in OOP

The methodology of predicting CR in OOP is the same as that for filtering. It is based on the scatterplot obtaining and curve fitting. The only difference is that the vertical axis relates to CR, while the horizontal axis, as earlier, corresponds to mean probability. Two mean probabilities $P_{2\sigma}$ and $P_{2.7\sigma}$ have been considered where the latter occurred to be worse again. Therefore, the obtained results for the mean probability $P_{2\sigma}$ only are presented below.

Two lossy compression methods, namely, the coders AGU and ADCT, have been studied. Their scatterplots are presented in Fig. 16. Contrary to other cases considered above, fitting is performed using a sum of two weighted exponential functions. As can be seen, fitting in both cases is very good with R^2 exceeding 0.99. Slightly larger values of CR are provided by the more sophisticated coder ADCT [60]. Very large (over 20) values of CR are provided for $P_{2\sigma} >$ 0.93, that is, for simple structure images corrupted by intensive noise.

We did not have real-life multichannel images corrupted by AWGN. But the hyperspectral data for the sensors Hyperion and airborne visible/infrared imaging spectrometer (AVIRIS) were available. Noise in them is signal dependent [14] with prevailing SD component for the model (22). The parameters of this noise were estimated in an automatic manner [11] and, thus, it became possible to apply VST (a generalized Anscombe transform with properly adjusted parameters) with converting noise into pure additive with unity variance.

Lossy compression in OOP neighborhood has been applied after VST. After decompression, inverse transform has to be applied, respectively. The obtained and predicted values of CR for Hyperion data are depicted in Fig. 17(a). As can be seen, the curves are in good agreement. There are some channels where predicted CRs are slightly larger than attained ones. This is

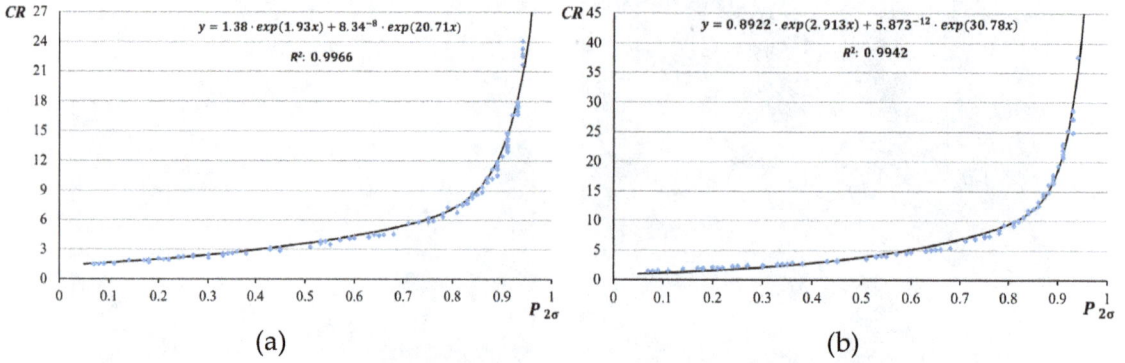

Figure 16. Scatterplots and fitted curves of dependences of CR vs. $P_{2\sigma}$ for the coders AGU (a) and ADCT (b)

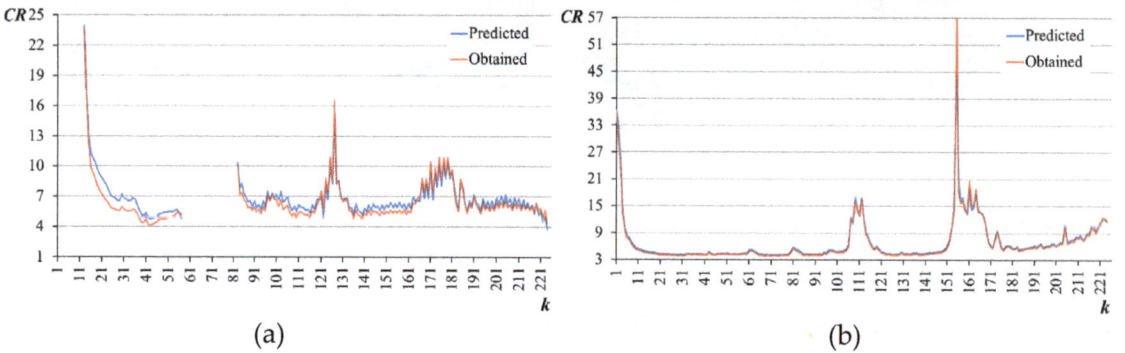

Figure 17. Results of a component-wise compression of Hyperion data (the analyzed set is EO1H1800252002116110KZ) (a) and AVIRIS Lunar Lake image (b) by the coder AGU after the generalized Anscombe transform

explained by the imperfectness of VST and blind estimation of noise parameters for channels with high signal-to-noise ratio. The largest CRs take place for sub-bands with low SNR (these are the sub-bands with indices 13–20, 125–130, and 175–180).

The results for the AVIRIS test image Lunar Lake are given in Fig. 17(b). Here, the agreement between the predicted and the attained values is even better than for the Hyperion data. Again, the largest CR is observed for sub-bands with low SNR. There are considerable differences in maximal and minimal values of CR. The main reason is the different SNR and different dynamic range in sub-band images. Certainly, CR also depends upon the image content.

5. Conclusions and future work

It is demonstrated that it is possible to predict the efficiency of image filtering as well as the parameters of lossy compression of a noisy image in OOP neighborhood. As opposed to the earlier known approaches that allow predicting potential efficiency of filtering, the present approach predicts practically a reachable performance and makes this very rapidly, by one or more orders faster than filtering or compression itself.

Certainly, a limited number of quality metrics, filtering, and compression techniques have been considered. However, it is important that a general methodology of prediction is proposed, and it is shown there are somewhat strict connections between simple input parameters (that can be easily and quickly calculated) and output parameters that are able to adequately characterize the efficiency of filtering or lossy compression techniques. In favor of this methodology, there are certain facts. First, there are many modern filters that have filtering efficiency of the same order as the DCT-based filter and BM3D. Thus, predicting denoising efficiency for the filters mentioned above, it is possible to approximately predict performance for other modern filters (although such prediction would be less accurate). Second, the same holds for lossy compression methods. For example, AGU and JPEG2000 provide similar performance characteristics. Then, by predicting compression parameters for AGU, they are, in fact, estimated for JPEG2000 as well.

Concerning the decision making, whether to perform filtering or not, strict recommendations have been given for probabilities $P_{2\sigma}$ and $P_{0.5\sigma}$. Filtering can be expedient if $P_{2\sigma}$ exceeds 0.5 or $P_{0.5\sigma}$ exceeds 0.15. Similarly, OOP is quite possible if $P_{2\sigma}$ is approximately 0.85 or larger. A very important fact is that these rules for filtering are valid for different types of noise (pure additive and signal-dependent, additive white Gaussian and spatially correlated). This generalization can be considered as one of the main contributions of this chapter. Meanwhile, the case of spatially correlated noise requires more attention in future. In prediction of filtering efficiency, general prediction approximations for spatially correlated noise with *a priori* known or pre-estimated properties (e.g., 2D spectrum) have not been obtained yet. It can only be expected that the scatterplots for spatially correlated noise with other (not analyzed yet) shapes and parameters of spatial power spectrum behave similarly. The studies for lossy compression of images corrupted by spatially correlated noise are yet to be started. This opens a very wide field for future research.

The results of this research show that although sometimes the prediction of performance characteristics based on one input parameter is appropriately accurate, there are several means to improve the prediction accuracy. One way that deals with multiparameter input has been already used for particular cases. The use of mean $P_{0.5\sigma}$ has shown itself a good solution, although it has not yet been tried for all possible applications. In particular, mean $P_{0.5\sigma}$ has not been tested for lossy compression. It is hoped that performance can be improved due to this reason. Neural networks or other approximators of multidimensional functions (surfaces) can be useful.

There are also other possible directions for future research. 3D filtering warrants a more thorough study, at least, for the case of more than three channels. The same relates to 3D lossy compression performance, which has not been tried to predict yet. Compression parameters for QS other than the one recommended for OOP is also of sufficient interest in DCT-based lossy compression. Influence of errors in *a priori* information on noise parameters or their blind estimates on prediction accuracy has to be studied as well.

Author details

Benoit Vozel[2*], Oleksiy Rubel[1], Alexander Zemliachenko[1], Sergey Abramov[1],
Sergey Krivenko[1], Ruslan Kozhemiakin[1], Vladimir Lukin[1] and Kacem Chehdi[2]

*Address all correspondence to: benoit.vozel@univ-rennes1.fr

1 National Aerospace University, Ukraine

2 University of Rennes 1, France

References

[1] Schowengerdt R. (2006). Remote Sensing, Third Edition. *Models and Methods for Image Processing*, Academic Press, Orlando, FL.

[2] Oliver C. & Quegan S. (2004). *Understanding Synthetic Aperture Radar Images.* SciTech Publishing, Herndon, VA.

[3] Christophe, E. (2011). Hyperspectral Data Compression Tradeoff. In: *Optical Remote Sensing in Advances, Signal Processing and Exploitation Techniques*, Eds. Prasad S., Bruce L. M., and Chanussot J., pp. 9-29. Springer.

[4] Lukin V., Abramov S., Ponomarenko N., Uss M., Zriakhov M., Vozel B., Chehdi K., & Astola J. (2011). Methods and automatic procedures for processing images based on blind evaluation of noise type and characteristics. *SPIE Journal on Advances in Remote Sensing*, Vol. 5, No. 1, 053502. Doi: 10.1117/1.3539768.

[5] Lukin V., Abramov S., Ponomarenko N., Krivenko S., Uss M., Vozel B., Chehdi K., Egiazarian K., & Astola J. (2014). Approaches to Automatic Data Processing in Hyperspectral Remote Sensing. *Telecommunications and Radio Engineering*, Vol. 73, No. 13, pp. 1125-1139.

[6] Chang C. I. (Ed.). (2007). *Hyperspectral Data Exploitation: Theory and Applications.* Wiley-Interscience, Hoboken, NJ.

[7] Aiazzi B., Alparone L., Barducci A., Baronti S., Marcoinni P., Pippi I., & Selva M. (2006). Noise modelling and estimation of hyperspectral data from airborne imaging spectrometers. *Annals of Geophysics*, Vol. 49, No. 1, pp. 1-9.

[8] Vozel B., Abramov S., Chehdi K., Lukin V., Ponomarenko N., Uss M., & Astola J. (2009). Blind methods for noise evaluation in multi-component images, In: *Multivariate Image Processing*, pp. 263-295. France.

[9] Ponomarenko N., Lukin V., Egiazarian K., & Lepisto L. (2013). Adaptive Visually Lossless JPEG-Based Color Image Compression. Signal, *Image and Video Processing*, Doi: 10.1007/s11760-013-0446-1, 16 p.

[10] Meola J., Eismann M. T., Moses R. L., & Ash J. N. (2011). Modeling and estimation of signal-dependent noise in hyperspectral imagery. *Applied Optics*, Vol. 50, No. 21, pp. 3829-3846.

[11] Uss M., Vozel B., Lukin V., & Chehdi K. (2011). Local signal-dependent noise variance estimation from hyperspectral textural images. *IEEE Journal of Selected Topics in Signal Processing*, Vol. 5, No. 2, pp. 469-486. Doi: 10.1109/JSTSP.2010.2104312.

[12] Uss M., Vozel B., Lukin V., & Chehdi K. (2012). Maximum likelihood estimation of spatially correlated signal-dependent noise in hyperspectral images. *Optical Engineering*, Vol. 51, No. 11. Doi: 10.1117/1.OE.51.11.111712.

[13] Abramov S., Zabrodina V., Lukin V., Vozel B., Chehdi K., & Astola J. (2011). Methods for Blind Estimation of the Variance of Mixed Noise and Their Performance Analysis. In: *Numerical Analysis – Theory and Applications*, Ed. J. Awrejcewicz, pp. 49-70. In-Tech, Austria, ISBN 978-953-307-389-7.

[14] Abramov S., Uss M., Abramova V., Lukin V., Vozel B., & Chehdi K. (2015). On Noise Properties in Hyperspectral Images. IGARSS, Milan, Italy, pp. 3501-3504.

[15] Zhong, P., Wang, R. (2013). Multiple-Spectral-Band CRFs for Denoising Junk Bands of Hyperspectral Imagery in *IEEE Transactions on Geoscience and Remote Sensing*, Vol. 51(4), pp. 2269-2275.

[16] Blanes I., Zabala A., Moré G., Pons X., & Serra-Sagristà J. (2009). Classification of hyperspectral images compressed through 3DJPEG2000. KES '08 Proceedings of the 12th International Conference on Knowledge-Based Intelligent Information and Engineering Systems, Part III, LNAI, Springer, Berlin, Heidelberg, Vol. 5179, pp. 416–423.

[17] Abramov S., Krivenko S., Roenko A., Lukin V., Djurovic I., & Chobanu M. (2013). Prediction of Filtering Efficiency for DCT-based Image Denoising. *Proceedings of MECO*, Budva, Montenegro, pp. 97-100.

[18] Pogrebnyak O. & Lukin V. (2012). Wiener DCT Based Image Filtering. *Journal of Electronic Imaging*, Vol. 4, No. 14, pp. 043020-043020.

[19] Zemliachenko A. N., Kozhemiakin R. A., Uss M. L., Abramov S. K., Ponomarenko N. N., Lukin V. V., Vozel B., & Chehdi K. (2014). Lossy compression of hyperspectral images based on noise parameters estimation and variance stabilizing transform. *Journal of Applied Remote Sensing*, Vol. 8, No. 1, 25 p. Doi: 10.1117/1.JRS.8.083571.

[20] Zemliachenko A., Abramov S., Lukin V., Vozel B., & Chehdi K. (2015). Compression Ratio Prediction in Lossy Compression of Noisy Images, Proceedings of IGARSS, Milan, Italy, pp. 3497-3500.

[21] Pyatykh S., Hesser J., & Zheng L. (2013). Image noise level estimation by principal component analysis. *IEEE Transactions on Image Processing*, Vol. 22, No. 2, pp. 687-699.

[22] Sendur L. & Selesnick I. W. (2002). Bivariate shrinkage with local variance estimation. *IEEE Signal Processing Letters*, Vol. 9, No. 12, pp. 438-441.

[23] Ponomarenko N. N., Lukin V. V., Egiazarian K. O., & Astola J. T. (2010). A method for blind estimation of spatially correlated noise characteristics. *Proceedings of SPIE 7532 of Image Processing: Algorithms and Systems VIII*, 753208, San Jose, USA, January 2010. Doi: 10.1117/12.847986.

[24] Lebrun M., Colom M., Buades A., & Morel J. M. (2012). Secrets of image denoising cuisine. *Acta Numerica*, Vol. 21, pp. 475-576.

[25] Van Zyl Marais I., Steyn W.H., & du Preez J.A. (2009). On-Board Image Quality Assessment for a Small Low Earth Orbit Satellite. *Proceedings of the 7th IAA Symp. on Small Satellites for Earth Observation*, Berlin, Germany.

[26] Anfinsen S. N., Doulgeris A. P., & Eltoft T. (2009). Estimation of the equivalent number of looks in polarimetric synthetic aperture radar imagery. *IEEE Transactions on Geoscience and Remote Sensing*, Vol. 47, No. 11, pp. 3795-3809.

[27] Liu C., Szeliski R., Kang S. B., Zitnick C. L., & Freeman W. T. (2008). Automatic estimation and removal of noise from a single image. *IEEE Transactions on Pattern Analysis and Machine Intelligence*, Vol. 30, No. 2, pp. 299-314.

[28] Colom M., Lebrun M., Buades A., & Morel J. M. (2014). A Non-Parametric Approach for the Estimation of Intensity-Frequency Dependent Noise. IEEE International Conference on Image Processing (ICIP). Doi: 10.1109/ICIP.2014.7025865.

[29] Mallat S. (1998). *A Wavelet Tour of Signal Processing*. Academic Press, San Diego.

[30] Öktem R., Yaroslavsky L., Egiazarian K., & Astola J. (2002). Transform domain approaches for image denoising. *Journal of Electronic Imaging*, Vol. 11, No. 2, pp. 149 – 156.

[31] Solbo S. & Eltoft T. (2004). Homomorphic wavelet-based statistical despeckling of SAR images. *IEEE Transactions on Geoscience and Remote Sensing*, Vol. GRS-42, No. 4, pp. 711-721.

[32] Öktem R., Egiazarian K., Lukin V.V., Ponomarenko N.N., & Tsymbal O.V. (2007). Locally adaptive DCT filtering for signal-dependent noise removal. *EURASIP Journal on Advances in Signal Processing*, Vol. 2007, 10 p.

[33] Lukin V., Ponomarenko N., Egiazarian K., & Astola J. (2008). Adaptive DCT-based filtering of images corrupted by spatially correlated noise. *Proceedings of SPIE 6812 of Image Processing: Algorithms and Systems VI*, 68120W, San Jose, USA. Doi: 10.1117/12.764893.

[34] Dabov K., Foi A., Katkovnik V., & Egiazarian K. (2007). Image denoising by sparse 3-D transform-domain collaborative filtering. *IEEE Transactions on Image Processing*, Vol. 16, No. 8, pp. 2080-2095.

[35] Bekhtin Yu. S. (2011). Adaptive Wavelet Codec for Noisy Image Compression. *Proceedings of the 9-th East-West Design and Test Symposium*, Sevastopol, Ukraine, Sept. 2011, pp. 184-188.

[36] Bazhyna A., Ponomarenko N., Egiazarian K., & Lukin V. (2007). Compression of noisy Bayer pattern color filter array images. *Proceedings of SPIE Photonics West Symposium*, San Jose, USA, Jan. 2007, Vol. 6498, 9 p.

[37] Makitalo M., Foi A., Fevralev D., & Lukin V. (2010). Denoising of single-look SAR images based on variance stabilization and non-local filters. *CD-ROM Proceedings of MMET*, Kiev, Ukraine, 4 p.

[38] Ponomarenko N., Silvestri F., Egiazarian K., Carli M., Astola J., & Lukin V. (2007). On Between-Coefficient Contrast Masking of DCT Basis Functions. *CD-ROM Proceedings of VPQM*, USA, 4 p.

[39] Lukin V., Abramov S., Krivenko S., Kurekin A., & Pogrebnyak O. (2013). Analysis of classification accuracy for pre-filtered multichannel remote sensing data. *Journal of Expert Systems with Applications*, Vol. 40, No. 16, pp. 6400-6411.

[40] Lukin V. & Bataeva E. (2012). Challenges in Pre-processing Multichannel Remote Sensing Terrain Images, Importance of GEO initiatives and Montenegrin capacities in this area. *The Montenegrin Academy of Sciences and Arts*, Book No. 119, The Section for Natural Sciences, Book No. 16, pp. 63-76.

[41] Al-Shaykh O. K. & Mersereau R. M. (1998). Lossy compression of noisy images. *IEEE Transactions on Image Processing*, Vol. 7, No. 12, pp. 1641-1652.

[42] Ponomarenko N. N., Lukin V. V., Egiazarian K., & Astola J. (2005). DCT Based High Quality Image Compression. *Proceedings of 14th Scandinavian Conference on Image Analysis*, Joensuu, Finland, Vol. 14, pp. 1177-1185.

[43] Zemliachenko A. N., Abramov S. K., Lukin V. V., Vozel B., & Chehdi K. (2014). Prediction of Optimal Operation Point Existence and Parameters in Lossy Compression of Noisy Images, *Proceedings of SPIE*, Vol. 9244, Image and Signal Processing for Remote Sensing XX, 92440H. Doi: 10.1117/12.2065947.

[44] Zemliachenko A., Abramov S., Lukin V., Vozel B., & Chehdi K. (2015). Lossy compression of noisy remote sensing images with prediction of optimal operation point existence and parameters. *SPIE Journal on Applied Remote Sensing*, Vol. 9, No. 1, pp. 095066-1-095066-26.

[45] Rubel A., Lukin V., & Egiazarian K. (2015). A method for predicting DCT-based denoising efficiency for grayscale images corrupted by AWGN and additive spatially correlated noise. *Proceedings of SPIE Symposium on Electronic Imaging*, SPIE, Vol. 9399, USA. Doi:10.1117/12.2082533.

[46] Dabov K., Foi A., Katkovnik V., & Egiazarian K. (2007). Color Image Denoising via Sparse 3D Collaborative Filtering with Grouping Constraint in Luminance-Chromi-

nance Space. *IEEE International Conference on Image Processing, ICIP,* Vol. 1, pp. 313-316.

[47] Lee J.S. (1983). Digital image smoothing and the sigma filter. *Computer Vision, Graphics, and Image Processing,* Vol. 24, No. 2, pp. 255-269.

[48] Chatterjee P. & Milanfar P. (2010). Is denoising dead? *IEEE Transactions on Image Processing,* Vol. 19, No. 4, pp. 895-911.

[49] Levin A. and Nadler B. (2011). Natural image denoising: Optimality and inherent bounds. *IEEE Conference on Computer Vision and Pattern Recognition (CVPR),* pp. 2833-2840.

[50] [50 Cameron C., Windmeijer A., Frank A.G., Gramajo H., Cane D.E., & Khosla C. (1997). An R-squared measure of goodness of fit for some common nonlinear regression models. *Journal of Econometrics,* Vol. 77, No. 2, pp. 1790–1792.

[51] Rubel O. & Lukin V. (2014) An Improved Prediction of DCT-Based Filters Efficiency Using Regression Analysis. *Information and Telecommunication Sciences,* Kiev, Ukraine, Vol. 5, No. 1, pp. 30-41.

[52] Badiru A. & Cheung J. (2002). Fuzzy Engineering Expert Systems with Neural Network Applications. Wiley-Interscience, New York.

[53] Rubel A., Naumenko A., & Lukin V. (2014). A Neural Network Based Predictor of Filtering. *Proceedings of MRRS,* Kiev, Ukraine, pp. 14-17.

[54] Krivenko S., Lukin V., Vozel B., & Chehdi K. (2014). Prediction of DCT-based Denoising Eficiency for Images Corrupted by Signal-Dependent Noise. *Proceedings of IEEE 34th International Scientific Conference Electronics and Nanotechnology,* Kiev, Ukraine, pp. 254-258.

[55] Lukin V., Abramov S., Rubel A., Naumenko A., Krivenko S., Vozel B., Chehdi K., Egiazarian K., & Astola J. (2014). An approach to prediction of signal-dependent noise removal efficiency by DCT-based filter. *Telecommunications and Radio Engineering,* Vol. 73, No. 18, pp. 1645-1659.

[56] Rubel A., Lukin V., & Pogrebnyak O. (2014). Efficiency of DCT-based denoising techniques applied to texture images, *Proceedings of Mexican Conference of Pattern Recognition,* Cancun, Mexico, pp. 261-270.

[57] Rubel O., Lukin V., & de Medeiros F.S. (2015). Prediction of Despeckling Efficiency of DCT-based Filters Applied to SAR Images, *Proceedings of 2015 International Conference on Distributed Computing in Sensor Systems,* Fortaleza, Brazil, pp. 159-168.

[58] Rubel O. S., Kozhemiakin R. O., Krivenko S. S., & Lukin V. V. (2015). A Method for Predicting Denoising Efficiency for Color Images. *Proceedings of 2015 IEEE 35th International Conference on Electronics and Nanotechnology (ELNANO),* Kiev, Ukraine, pp. 304-309.

[59] Ponomarenko N., Ieremeiev O., Lukin V., Egiazarian K., & Carli M. (2011). Modified Image Visual Quality Metrics for Contrast Change and Mean Shift Accounting. *Proceedings of CADSM*, Ukraine, pp. 305 - 311.

[60] Ponomarenko N., Lukin V., Egiazarian K., & Astola J. (2008). ADCT: A New High Quality DCT Based Coder for Lossy Image Compression. CD-ROM Proceedings of LNLA, Switzerland, 6 p.

3

Remote Sensing of the Glacial Environment Influenced by Climate Change

Arshad Ashraf, Manshad Rustam, Shaista Ijaz Khan,
Muhammad Adnan and Rozina Naz

Additional information is available at the end of the chapter

Abstract

Remote sensing-based observations prove to be critical for the monitoring and assessment of cryosphere in the Himalayan region, where routine data collection in mountainous regions is often hampered by highly inaccessible terrain and harsh climatic conditions. The glacierized region of High Asia is also facing the effects of climate change in the form of rapid melting of glacial ice, creation of new lakes, and expansion of the existing ones, which eventually result in hazardous glacial floods downstream. Multisensor remote sensing (RS) data, e.g., MODIS, Landsat-7 & 8, and SPOT-5 XS, coupled with Google Earth and digital elevation model (DEM) data were used to investigate the snow/glacier resources and their dynamics in the Karakoram–Himalaya basins adopting variable image interpretation and modeling techniques. Minimum numbers of large-sized glaciers were identified in the Himalaya range, which points toward higher rates of glacial ice melting in this range. On the contrary, the presence of relatively higher numbers of medium- to large-sized glaciers in the Karakoram range provides an evidence of favorable climate conditions for the glaciers' existence at higher altitudes. A significant gain in snow cover was observed in Hunza basin during the 2001–2011 period, which may feed high-altitude zone resulting in net expansion of the snow cover and ice mass gain in the Karakoram. The integrated use of RS and geographical information systems (GIS) techniques with sparse in situ data is found to be helpful in analyzing the glacial environment in the context of changing climate in the high-altitude Himalayan region.

Keywords: Snow cover, Glacial environment, Climate change, Karakoram range

1. Introduction

The glacierized region of Hindu Kush–Karakoram Himalaya (HKH) often referred to as the 'water tower of Asia' stores large volumes of water in the form of ice and snow after the polar ice releases freshwater to the Indus, Ganga, and Brahmaputra rivers. Climate change is being predicted by glacial lakes due to their property of acting as sensitive indicators [1], and unstable lakes can pose potential threats to downstream communities and infrastructure [2]. Monitoring of glaciers, glacial lakes, and assessment of glacial lake outburst flood (GLOF) impact downstream can be done quickly and rather reliably through remote sensing data interpretation and analysis. RS technology and GIS have often been used by decision makers as an effective and powerful tool to solve environmental issues [3]. In combination with GIS, RS methods provide useful means to detect potentially hazardous situations and to perform a preliminary assessment of the related hazard potential [4]. Remote sensing data from satellites are very helpful for mapping and monitoring glaciers and their changes over large areas, repeatedly, and by covering large regions with sufficient spatial detail at the same time [5, 6]. In combination with digital elevation models, RS data and methods offer the possibility to generate standardized glacier and glacial lake inventories.

When electromagnetic (EM) energy encounters matter, such as solid, liquid, or gas, a number of interactions are possible that may take place at the surface or beneath the surface of a substance. These interactions produce numerous changes in the incident EM radiation primarily in the form of change in magnitude, direction, wavelength, polarization, and phase. The science of RS detects and records these changes, which can be interpreted to identify the characteristics of the matter or land use/land cover, such as various types of vegetation cover, water bodies, soils, farming fields, and exposed rocks. The reflectance characteristics of the features like snow and ice vary according to their surficial/physical characteristics: the reflectance of snow is generally very high in the visible portions of the spectrum, whereas the reflectance of old snow and ice is always lower, i.e., due to compaction and presence of impurities, than that of fresh snow and clean/fresh glacier [7]. Similarly, the reflectance of fine-grain snow is comparatively higher than that of coarse-grain snow and glacial ice in the visible portion of the spectrum [8] (Figure 1). Comprehensive reviews of remote sensing systems, data types, techniques, and application to glacier-related hazards have been provided in Refs. [9–11]. RS technique provides additional opportunities for more complete surveys of glaciers to provide early warning of the potential formation of ice-dammed lakes [12]. The widely used earth observation (EO) sensors in the context of glacier inventory production include the Landsat MSS (Multispectral Scanner), TM (Thematic Mapper), ETM+ (Enhanced Thematic Mapper Plus), OLI (Operational Land Imager), ASTER (Advanced Spaceborne Thermal Emission and Reflection Radiometer) onboard the Terra platform, and the SPOT (Satellite Pour l'Observation de la Terre) satellites. In combination with freely available DEM datasets, remote sensing data offer integrative approaches for observing and assessing the current situation of glaciers, glacier lakes, and associated hazard potential, as well as the means to develop scenarios of potential future evolutions [13]. The snow and ice-melt model like SRM (snowmelt runoff model), which is used to simulate and forecast daily streamflow in snowy and glacierized basins [14], requires accurate data on snow-cover area (SCA), which are provided by Landsat, Terra-MODIS (Moderate Resolution Imaging Spectroradiometer), ERS-SAR (European Remote Sensing-Synthetic Aperture Radar), and NOAA-AVHRR (National Oceanic and Atmospheric Administration-Advanced Very High Resolution Radiometer) satellite sensors

[15]. For the first time, NOAA began to use remote sensing in 1966 for the detection of SCA to provide weekly estimates of snow cover in the Northern Hemisphere [16].

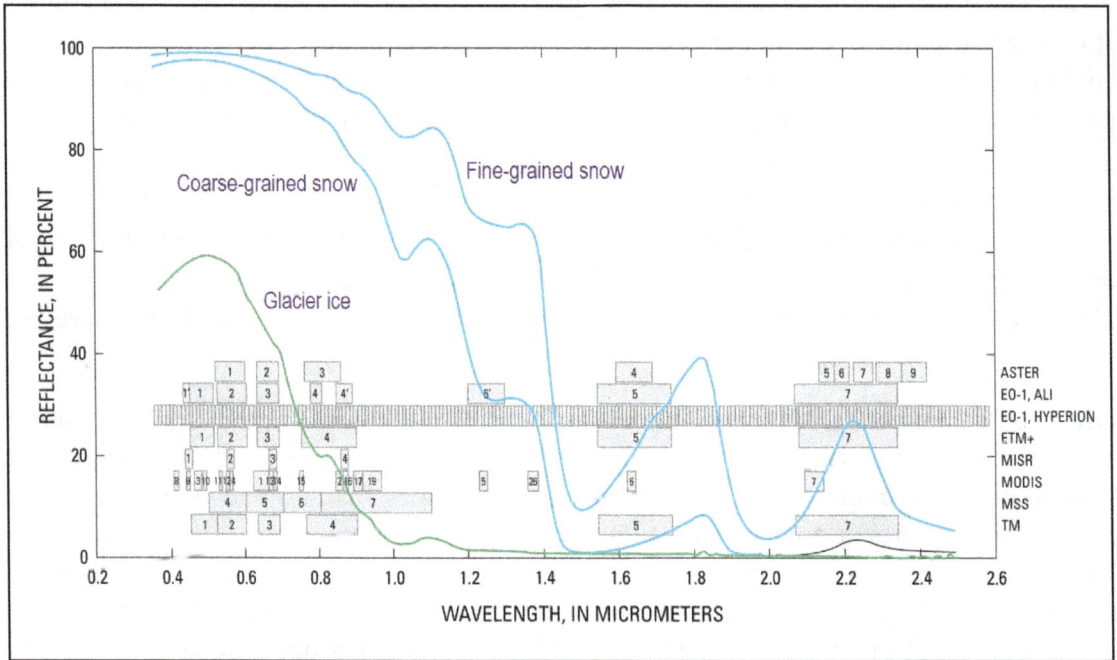

Figure 1. Reflectance response of snow and glacial ice in multisensor remote sensing data [8].

2. Climate change impacts on glacial environment

Glaciers are considered as very reliable and easily understandable natural indicators of climate change [17] due to their sensitive response to changes in temperature and precipitation [18]. They have been selected for this reason as an essential climate variable (ECV) by the global climate observing system (GCOS) [19]. According to IPCC [20], the global temperature has risen by 0.85°C since 1880 and the surface warming amounting to 3.7°C will be likely between 2081 and 2100 if greenhouse gas emissions stay roughly on their current path. The observed and projected changes in global average temperature relative to the 1986–2005 average under four emissions pathways are shown in Figure 2. The increase in air temperature influences the glacier mass balance [21], which is the balance between accumulation and ablation of glaciers [22]. The changes in mass balance cause variations in the volume and thickness of glaciers, which ultimately affect the flow of ice [21], and as a large fraction of the Indus flow is originated from meltwater, both magnitude and timing of the flow are vulnerable to climate change [23]. Furthermore, due to the temperature increase in the region, more precipitation in winter will fall as rainfall than as snowfall compared with the current situation. This rainfall will be added directly to the river system, instead of storing in the form of ice or snow in glaciers [24].

The IPCC Report of 2007 estimates a further warming of 3.7°C at the end of the 21st century; climate change has been observed through significant warming in the Hindu Kush Himalayas [25]. The climatic change in recent decades has made considerable impact on the glacier life cycle in the Himalayan region. With few exceptions, there has been a global trend toward glacier retreat since the beginning of the 20th century, with this retreat becoming more rapid and more uniform since the 1980s [26]. There will be a decrease of the glacier coverage in the coming decades as a result of global warming. This will lead to a short-term increase in water availability, in the coming decades, due to an increase in meltwater. However, the water availability will decrease in the long term during the second half of the 21st century. This decrease in water availability combined with a projected increase in water demand will cause water shortage for irrigation and thus food insecurity [27].

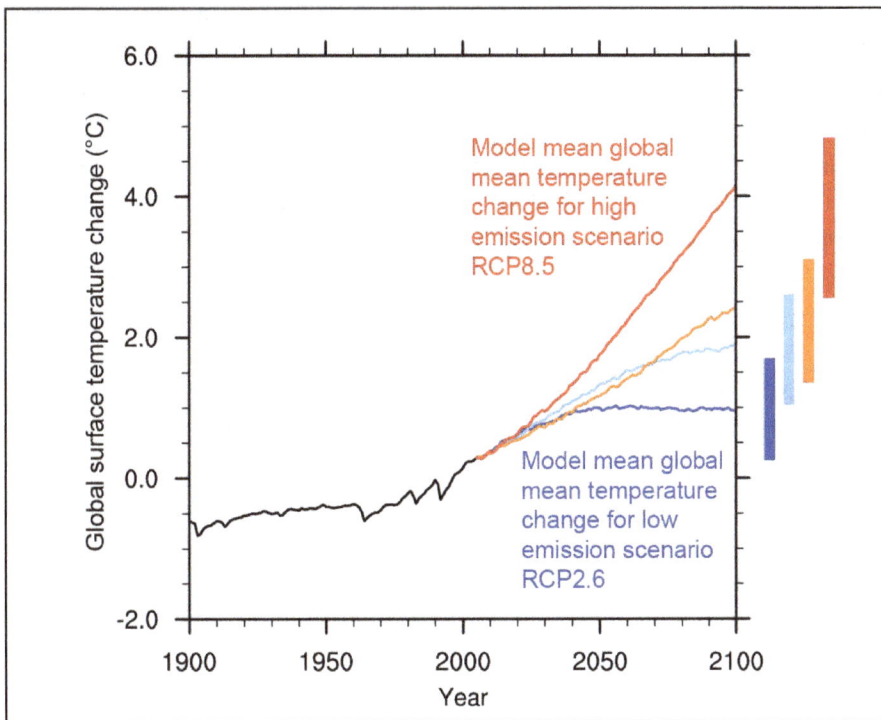

Figure 2. The changes in observed and projected global average temperature relative to the 1986–2005 average. The projections are averaged across a range of climate models. The vertical bars shown at right are likely ranges in temperature by the end of the century [20].

Some of the serious consequences of global warming in the Himalayan region include rapid melting of glaciers, creation of new lakes, and expansion of the existing ones posing high risk of glacial lake outburst flood hazard for downstream communities. The sudden increase in the frequency of floods in recent years, e.g., during 2007, 2008, 2010, 2012, and 2013 [28], demands a better understanding and investigation of the prevailing situation of the glaciers and glacial lakes in this region. The chapter describes a remote sensing-based approach to investigate environmental challenges posed by global warming in the Himalayan region.

2.1. Case study

According to Chaudhry *et al.* [29], Pakistan experienced 0.76°C rise in temperature during the last four decades and the increase was 1.5°C in the mountain environment hosting thousands of glaciers. The average annual temperature and annual rainfall at Gilgit meteorological station in the Central Karakoram indicated overall rising trends during the 1960–2013 period (Figure 3). Under varying climate conditions, glaciers in various regions of the Hindu Kush–Karakoram–Himalayan belt behave differently under changeable climate conditions. A general shrinkage of glaciers has been observed in the Himalaya [30, 31]; however, this does not imply a synchronous behavior of all glaciers, because there can be local differences and even advancing of existing glaciers [32, 33]. In the present study, snow-cover mapping of Hunza River basin situated in the Karakoram range of Pakistan was carried out using MODIS snow product for assessment of snow-cover dynamics under the changing climate. Multisensor RS data, i.e., MODIS product, LANDSAT-7 ETM+ (Enhanced Thematic Mapper plus), LANDSAT-8, and SPOT-5 XS (Multispectral) coupled with Google Earth and digital elevation model data (ASTER/SRTM) were used to investigate the snow/glacier resources and their dynamics in the selected Karakoram and Himalayan basins adopting variable image interpretation and modeling techniques. The snowmelt runoff model was employed to simulate the daily discharges at Gilgit stream gauging station in Gilgit River basin. World Meteorological Organization (WMO) tested SRM successfully for runoff simulations [34]. The model has been applied widely all over the world to compute snowmelt runoff. With the development of satellite remote sensing (SRS) and GIS, it is possible to apply SRM to a large-sized basin. It uses remote sensing snow-cover data for the estimation of snowmelt runoff. The study would provide base for future monitoring of glaciers and glacial lakes in response to changing climate in this high-altitudinal mountainous region.

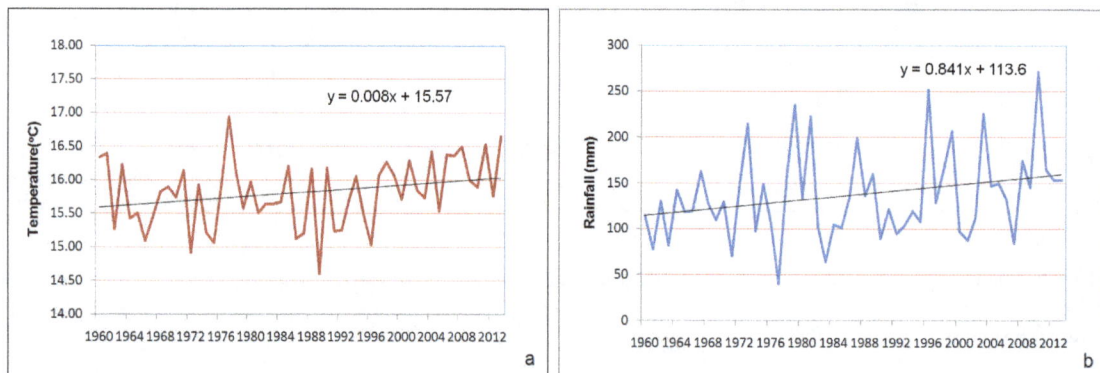

Figure 3. Trends of average annual temperature (a) and annual rainfall at Gilgit (b) during 1960–2013.

2.2. Description of the study area

The glacierized region of Pakistan lies within longitudes 70° 57′–77° 52′ E and latitudes 33° 52 ′–37° 09′ N (Figure 4). The elevation ranges from 366 m in the south to more than 8,500 m in the northeast. The snow and glacial ice reserves of freshwater nourish the main Indus River system (IRS) of the country. Approximately 11.57% of the overall area (i.e., 22,000 km²) of Upper Indus basin (UIB) is covered by seasonal glacial ice occupied by majority of the largest valley glaciers, the biggest and prevailing snow/ice-covered area outside the polar regions [35]. The high mountain region, i.e., between 35° and 37° N, is mostly dominated by winter rains, whereas the submountainous region, i.e., between 33.5° and 35°N, is dominated by summer rains. The bulk of the snowfall received from westerlies during the winter half of the year and more local conditions prevail in winter under the existing influence of the Tibetan anticyclone [36]. In addition to the influence of global weather systems, the mountain climates are also influenced on the medium and local scale by elevation, valley orientation, aspect, and slope [37]. The Himalayas have four subregions. The sub-Himalayas or Siwaliks are a range of low hills up to 1,000 m altitude above the mean sea level. The outer Himalayas go up to about 5,000 m altitude. The central Himalayas have an average height of about 6,000 m. The trans-Himalayas including the Karakoram Range are also very high, which include the second highest peak (8,611 m) in the world. The Hindu Kush and the Western Mountains form the boundary between Pakistan, Afghanistan, and China. The main rivers in these ranges are Swat and Kabul, which eventually run into the river Indus.

Gilgit basin is bounded in the west by Chitral River basin, a small portion in the north by Afghanistan, in the east by Hunza River basin, and in the south by Indus and Swat River basins. The basin occupies an area of about 14,082.4 km² out of which about 6.9% is glacierized. The elevation ranges from 1,500 masl to more than 6,500 masl. Hunza basin is located in the upstream part of Upper Indus basin covering an area of about 14,235 km² in which about 27.6% area is glacierized. The Hunza River has formed the main subbasin of the Gilgit basin. The tributaries joining the Hunza River are Chapursan, Khunjerab, Ghujerab, Shimshal, and Hispar rivers. Generally, most parts of the ablation areas are debris covered in this region. The Hunza River gauged at Dainyor bridge has a mean annual flow of 323 m³ s⁻¹ based on 1966–2008 flow record of the Surface Water Hydrology Project of the Water and Power Development Authority (SWHP-WAPDA). The Astor basin lies in the eastern side of the Nanga Parbat mountain. Astor River drains the snow- and glacier-covered mountains of Ladakh – Deosai and High Himalayas in the northern territory of Pakistan. Shingo basin (4,680 km²) lies in the southeast of Astore basin within the elevation range of 3,800–6,000 m. Generally, the glaciers are few in number and small in size in this basin. The Jhelum basin is bounded in the west by southwestern part of Indus River basin, in the north by Astore basin, and in the east by Shingo basin (Figure 4). The elevation in this basin ranges from 1,200 m to more than 4,700 m.

3. Material and methods

3.1. Data used

A dataset of MODIS processed images of MOD10A2 (h23V05, h24V05) and MYD10A2 (h23V05, h24V05) available since 2000–2011 was downloaded from the web link http://nsidc.org/cgi-bin/snowi/ with a minimum cloud cover of 15%. MODIS is an optical sensor that

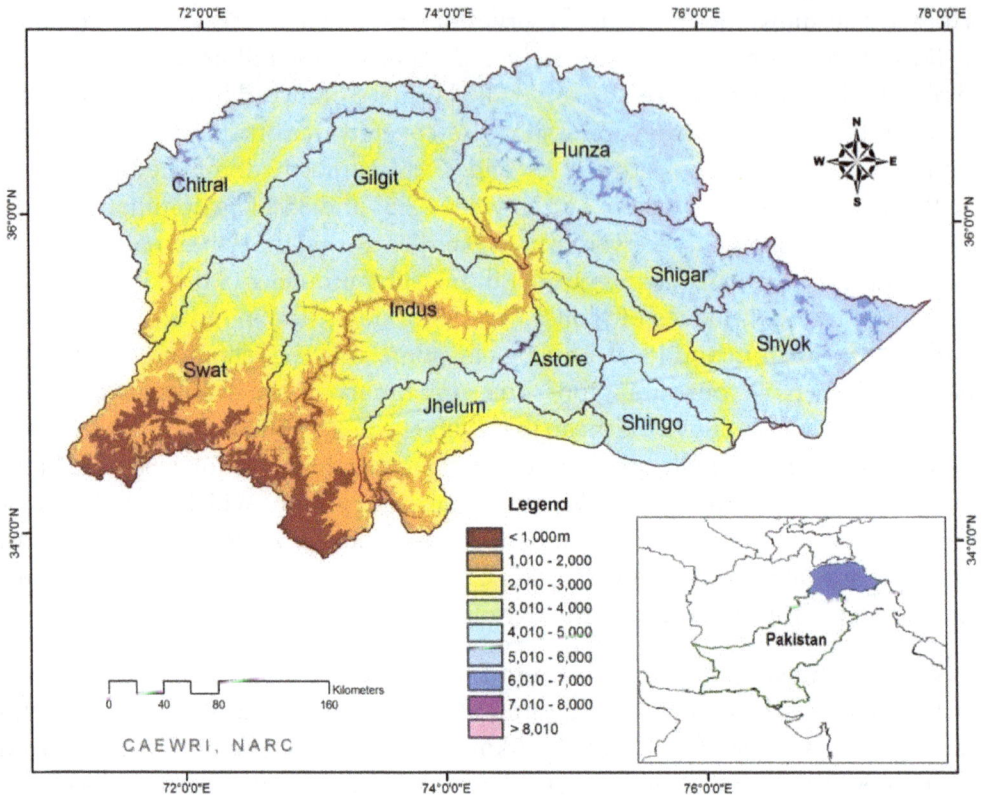

Figure 4. Location map of the glacierized region of Pakistan indicating various river basins and altitudinal ranges.

provides imagery of the earth's surface and clouds in 36 discrete, narrow spectral bands ranging from 0.4 to 14.4 µm of the electromagnetic spectrum. MODIS snow-cover images are available globally at a variety of different resolutions and projections. MODIS, aboard terra spacecraft of earth observing systems (EOS), is being very handy for the estimation of normalized differential snow index (NDSI). The MODIS snow-cover product used in this study (MOD10A2 and MYD10A2) contains data fields for maximum snow extent over an 8-day repeated period and has a spatial resolution of 500 m covering the Hunza River basin completely in two scenes (h23V05 and h24V05).

The glaciers and glacial lakes mapping was based on the Landsat 7 ETM plus and Landsat 8 satellite data of 2001 and 2013, respectively. The later data were downloaded from the web link http://glovis.usgs.gov with minimum cloud and snow cover. The detail of satellite data used in the present study is given in Table 1. The RS analysis for glacial lakes mapping was supplemented by Google Earth imageries and the topographic maps published by Survey of Pakistan. The Landsat 8 satellite images the entire earth every 16 days in an 8-day offset from Landsat 7 ETM plus. The images are terrain-corrected having spatial resolution of 30 m for multispectral bands 1–7, 9, and 15 m for panchromatic band 8 and 100 m for thermal infrared sensor (TIRS) bands 10–11 resampled to 30 m to match the multispectral bands. The Landsat 8 carries two instruments: the OLI sensor includes refined heritage bands, along with three new bands and thermal infrared sensor provides two thermal bands. The satellite remote sensing data of 1993 (Landsat MSS), 2001 (Landsat TM), and 2005 (SPOT XS multispectral) periods acquired from various sources, such as SUPARCO, were used for spatiotemporal analysis of glaciers and lakes in the Astore basin. The location of selected glaciers and lakes in the basin is shown in Figure 5. For historical trend analysis, glacier cover from topographic map of 1:50,000 scale of the year 1964 was acquired from Survey of Pakistan. Digital elevation model data of shuttle radar topography mission (SRTM) 90 m were used to estimate altitudinal characteristics of the glacial lakes. The DEM is provided with a geographic coordinate system (CGS), and the elevation values refer to datum WGS-84 both horizontally and vertically.

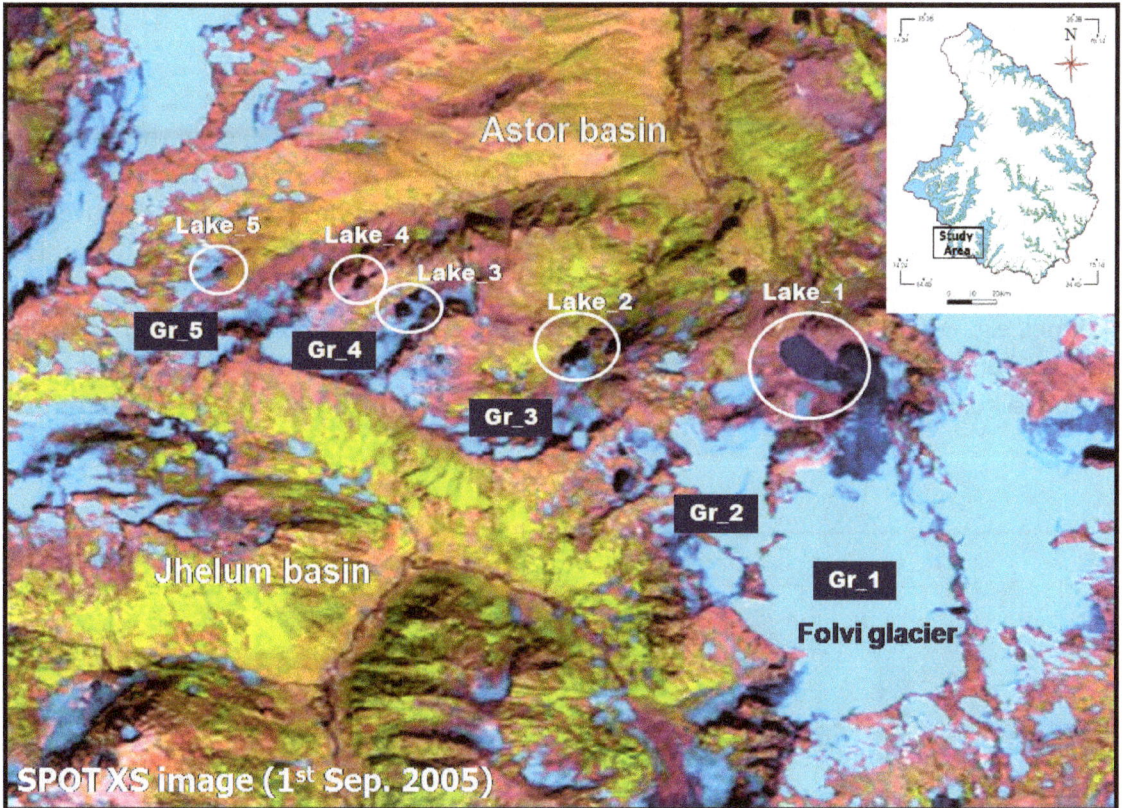

Figure 5. Location of the study site in Astore basin in Himalayas.

The daily flows data of Hunza and Astore rivers were acquired from SWHP-WAPDA for seasonal correlation with climate data and snow-cover dynamics in the catchment since 2000. The Hunza River is gauged at Dainyor Bridge, whereas the Astore River is gauged at Doyean station near Bunji.

3.2. Image processing and geo-spatial analysis

Originally downloaded MODIS product was in sinusoidal projection, which was then re-projected into Universal Transverse Mercator (UTM) Zone 43N projection with datum WGS-1984 using MODIS Re-projection Tool. MODIS images only for the months of January,

SRS data	Resolution	Period
MODIS snow-cover product	500 m	2000–2011
Landsat 8 OLI/TIRS	30 m	2013
Landsat 7 ETM+ (Enhanced Thematic Mapper Plus)	15 m, 30 m	2001
Landsat MSS (Multispectral Scanner)	80 m	1993
SPOT 4 XS (multispectral)	20 m	2005
Google Earth	Variable	Variable

Table 1. Satellite remote sensing data used in the present study.

February, and March were mosaicked and used for mapping the maximum SCA to observe snow-cover dynamics in Hunza basin, Karakoram range. The snow cover in these months usually dominates most of the basin area. The month-wise distribution of SCA in the study area during 2011 is shown in Figure 6. The SCA was minimum during August, followed by July, June, and May. It appears to increase from September to March then starts declining. The maximum snow-cover area change (SCAC) was assessed using the MODIS snow-cover product, e.g., MOD10A2 and MYD10A2, available since 2000. Subset of the study area was masked from the MODIS layer. The snow extracts of maximum snowfall period consist of the five classes of the MODIS attribute data from which the snow class (value = 200) was extracted.

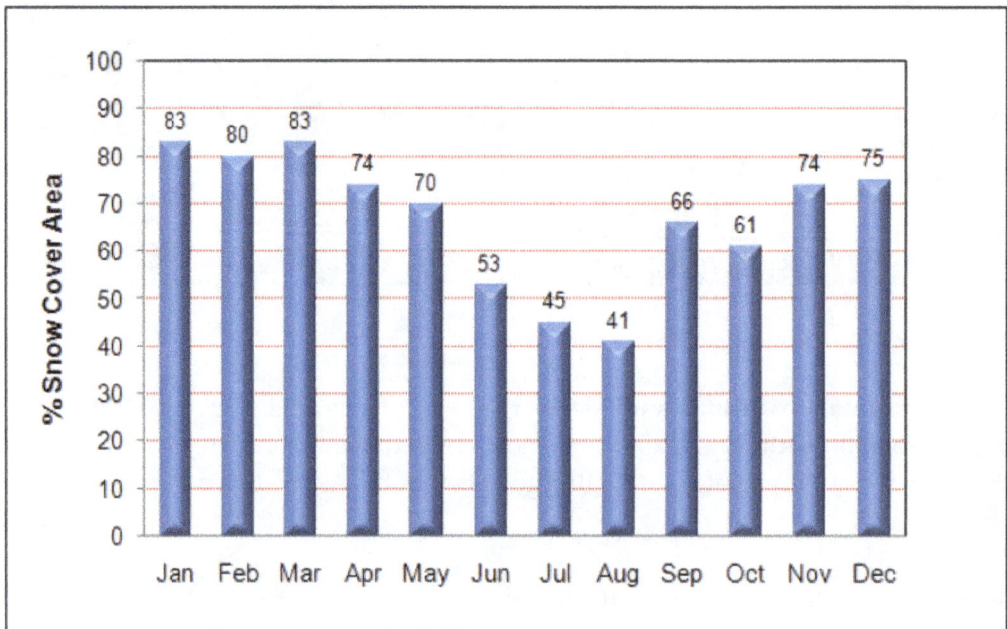

Figure 6. Monthly distribution of maximum SCA during 2011 in Hunza basin.

The spatiotemporal and altitudinal changes in the lakes were studied to observe the influence of climatic changes occurred during recent decades in this part of the HKH region. The spatial database of the lakes such as location coordinates, area, and length was systematically developed and analysis was performed for each river basin for 2001 and 2013. For the glacial lakes mapping, the methodology developed by Lanzhou Institute of Glaciology and Geocryology, the Water and Energy Commission Secretariat, and the Nepal Electricity Authority [38] was adopted. The uncertainty analysis for lakes area was performed following the methods provided by researchers, e.g., Refs. [39, 40]. According to the analysis, the shoreline of the glacial lake passes through the center of pixel giving an uncertainty of 0.5 pixel.

Five glaciers and five associated glacial lakes were selected in Astore basin of the Himalaya range. Spatial data layers of the glaciers and glacial lakes were developed through on-screen digitization in GIS and using different analytical techniques and logical operators. All the polygons representing glaciers and glacial lakes are numbered clockwise sequentially. For geospatial analysis, the attribute data were linked to the spatial data layers of glaciers and glacial lakes in GIS. Time series data of hydrometeorology were used to study the trends in climate data, i.e., summer and winter temperatures (maximum and minimum), precipitation, and river discharge.

3.3. Remote sensing technique in glaciers and lakes mapping

The detection of glacial lakes using multispectral imagery involves discriminating between water and other surface types. Delineating surface water can be achieved using the spectral reflectance differences. Water strongly absorbs in the near- and middle-infrared wavelengths (0.8–2.5 μm). Vegetation and soil, in contrast, have higher reflectance in the near- and middle-infrared wavelengths; hence, water bodies appear dark compared with their surroundings when using these wavelengths [41]. Methods for semiautomated mapping of glaciers and lakes based on remote sensing data have been well established for several years, and model approaches to assess the hazard potential of glacier lakes have been developed and successfully tested as well. The global elevation datasets of the shuttle radar topography mission and the ASTER global DEM (GDEM) offer the possibility to derive such topographic parameters for glaciers in most regions of the world.

The spatial and radiometric resolution of panchromatic band of Landsat ETM plus images was used for delineation of glacier's boundaries in selected basins. The very low reflectance of ice and snow in the middle-infrared has been widely used for glacier classification, for example, with threshold ratio images from raw data of digital number of TM bands 4 and 5 [42, 43]. This technique has proved to be simple, robust, and accurate [44]. It has also been proposed as a method for reducing multiple effects (e.g., the topographic effect of direct light) within multispectral data [45–47]. Visual interpretation is considered to be the most accurate way to delineate snow line in the scale of one outlet glacier, because it is the only method to take topography into account [48]. Because of a pronounced topographic effect, none of the most common band ratios or principal components could offer sufficient contrast to set one threshold value to delineate the ice-cap glaciers. The glaciers were mapped based on methodology developed by the Temporary Technical Secretary (TTS) at the Swiss Federal Institute of Technology, Zurich, for the data compilation of World Glacier Inventory [49]. The flowchart of the methodology adopted is shown in Figure 7. After digitization of glaciers' polygons, the numbering of glaciers was started from mouth of the major stream and proceeded clockwise round the basin. If there is a name assigned to the glacier, it was recorded through literature search and information included in the topographic maps. The geographic location of the glacier was recorded from the grid. The area of the glacier was calculated through database of the delineated glacier.

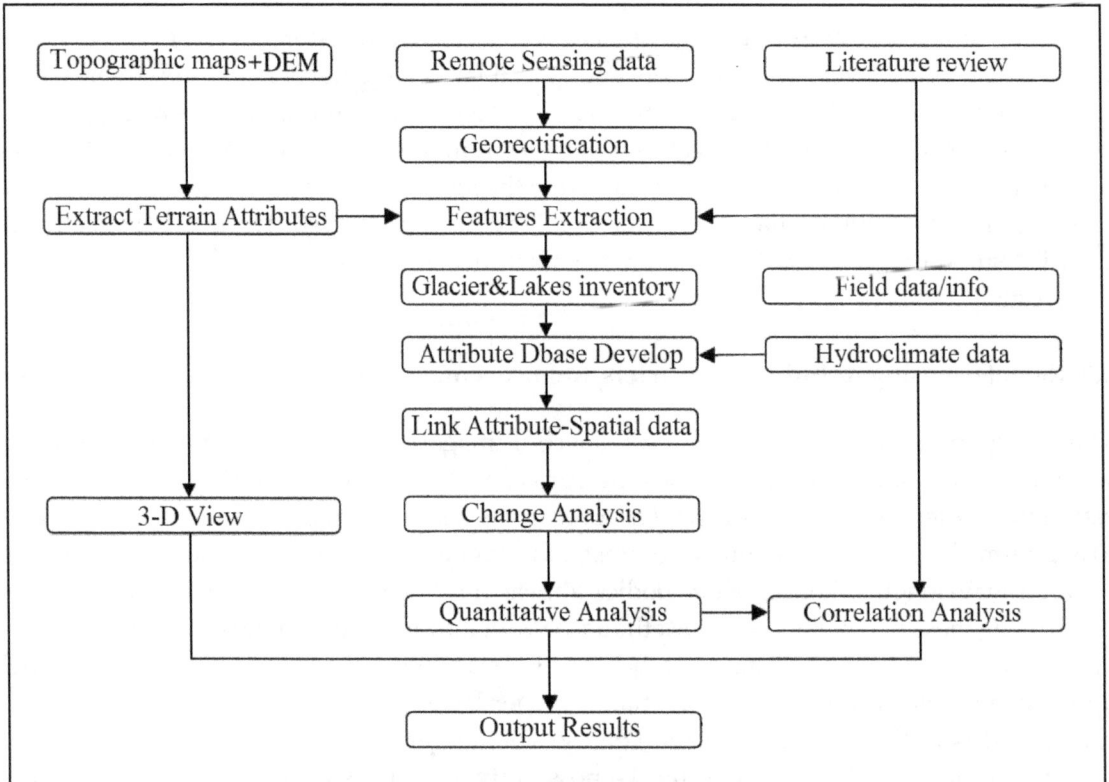

Figure 7. Flowchart of methodology adopted for temporal analysis of glaciers and glacial lakes.

3.4. Snow Runoff modeling

Currently, SRM is being used to analyze the effects of changed climate on seasonal river flows in snow and glacierized basins using MODIS satellite data. The daily input data used in the model are precipitation, air temperature, and snow-covered area. The model also requires some basin characteristics such as latitude–longitude, number of zone, zone areas, and means hypsometric elevation of each zone of basin. In the model phenomena, snowmelt and rain are computed every day and then superimposed on the calculated recession flow and transformed into the daily discharge from the catchment. The main equation used in SRM for snowmelt runoff simulation is:

$$Q_{n+1} = \left[c_{Sn} a_n \left(T_n + \Delta T_n \right) S_n + c_{Rn} P_n \right] \frac{A \times 10{,}000}{86{,}400} \left(1 - k_{n+1}\right) + Q_n k_{n+1} \tag{1}$$

where Q is the average daily discharge (m³/s), C_{sn} and C_{Rn} are the coefficients of snow and rain, respectively, a_n is the degree-day factor (cm °C⁻¹d⁻¹), T_n is the number of degree days in °C d, S is the ratio of the snow-covered area to the total area, P is the precipitation contributing to runoff (cm), T_{crit} (°C) is the critical temperature that differentiates between snow and rain, A is the area of the basin or zone in km, K is the recession coefficients that indicate the decline of discharge in a period without snowmelt or rainfall, and n is the sequence of days during discharge computation period. The degree-day factor is evaluated with regard to snow density, stage of the snowmelt season, and presence of glaciers. The runoff coefficient is an expression of hydrological losses and is estimated by comparing the annual precipitation and runoff, by taking into account the vegetation and current snow coverage, as well as size of the basin. The critical temperature (T_{crit}) can be estimated using actual meteorological records, stage of snowmelt season, and visual observations.

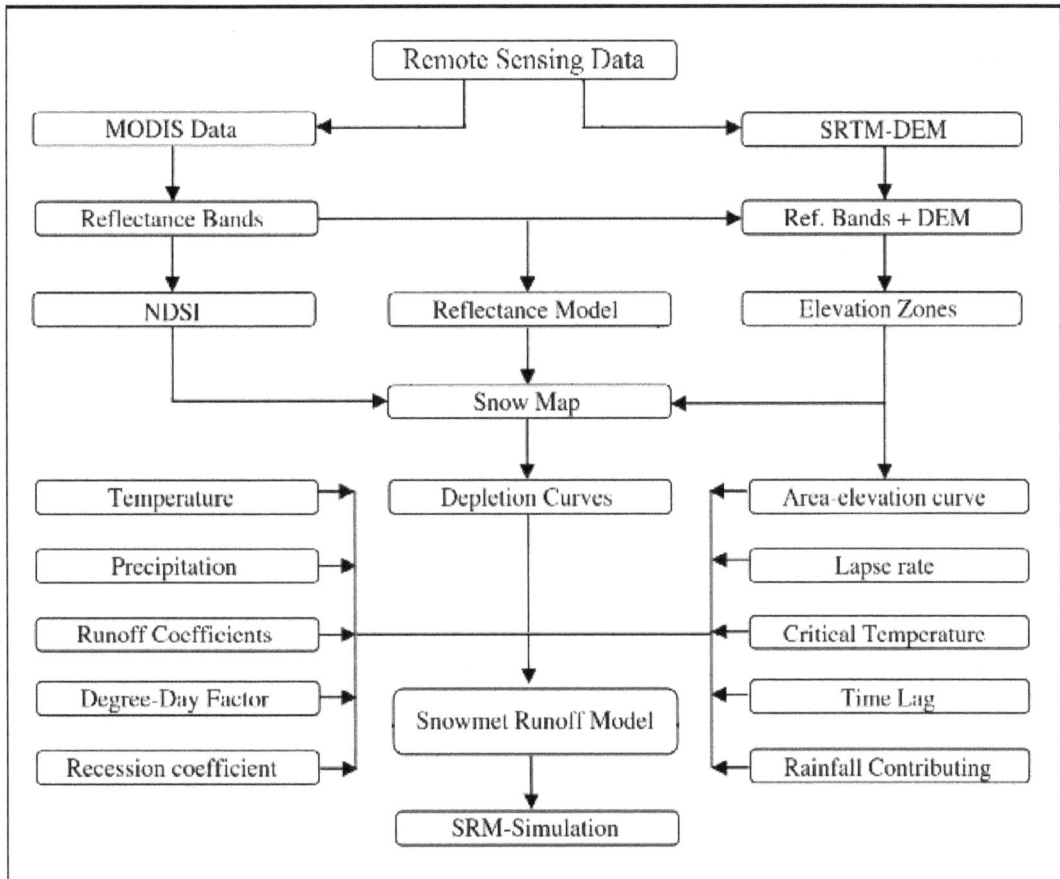

Figure 8. Major steps involved in simulation of snowmelt runoff model (SRM).

The Gilgit River basin is a snow-covered and glacierized basin, therefore the snowmelt runoff model can be successfully used to simulate and forecast daily stream flows as well as to study the effect of climate change on river flows. The SRM model was calibrated for 4 years from 2001 to 2004 and the model simulations were performed from year 2007 to 2010. The initial values of the parameters used during model calibration such as temperature lapse rate, degree-day factor snow and rain coefficients, and recession coefficients were extracted from past data and from previous studies, e.g., Ref. [50]. The parameters were adjusted during the calibration process until satisfactory results were achieved. The SRM model was calibrated with a coefficient of determination (R^2) value of 0.64 and validated with R^2 value of 0.78, indicating a close agreement between the observed and the simulated discharge data. The stepwise methodology followed in the study is shown in Figure 8. Different scenarios were used in SRM to predict future flows of Gilgit River: i. under rise in annual temperature 'T' and ii. increase in cryosphere area in the basin.

4. Results and discussion

4.1. Analysis of maximum snow-cover area

The maximum SCA in the Hunza basin, Karakoram range, was evaluated for trend and change analysis using MODIS product of the 2001–2011 period. Figure 9 shows the maximum snow-cover area in the Hunza basin during the 2001–2011 period. Except central valleys consisting of drainage network of the basin, most of the land appears to be snow covered during the period from January to March. From the results, it was observed that percentage SCA is predominantly increasing with time in this high-altitude cryospheric region. From the trend analysis of percentage snow-cover area on annual basis, it was observed that more than 80% of the basin area was snow covered particularly during the time periods of 2003, 2004, 2005, 2009, and 2011 (Figure 10). The maximum SCA during these years ranged from 80% to 92% of the study area. The historical data of precipitation (1961–2000) exhibited a rising trend in the Northern areas [51]. This may give rise to an increasing trend of SCA, which likely feeds the high-altitude zones (usually above 5,000 masl), resulting in net expansion of the snow cover and ice mass gain in the basin.

The maximum snow-cover area change observed during 2001–2011 indicated a significant increase of about 719 km^2 in SCA during an 11-year period (Figure 11). From the trend analysis of maximum snow extent on annual basis, it has been observed that the maximum snowfall month has shifted slightly from December to January indicating a spatial shift of winter season, which generally starts from November and continues until the end of March. The possible reasoning for this shift might be attributed to the observed facts that the solid precipitation in winter has been converted into liquid precipitation probably due to increased atmospheric temperatures.

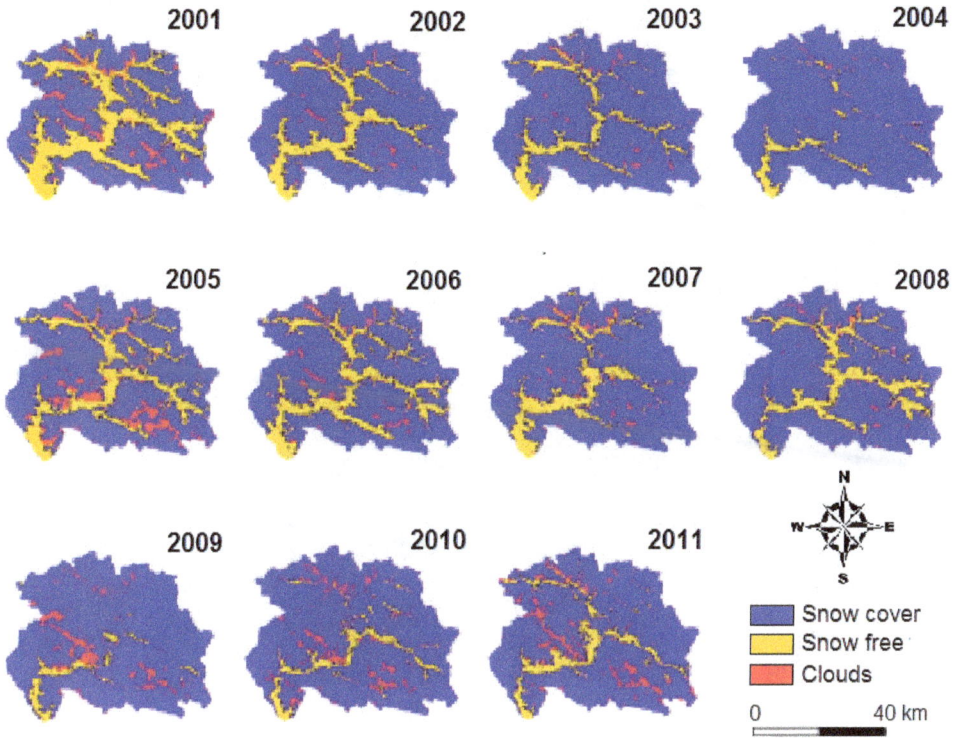

Figure 9. Maximum snow-cover area (SCA) in Hunza basin during the 2001–2011 period.

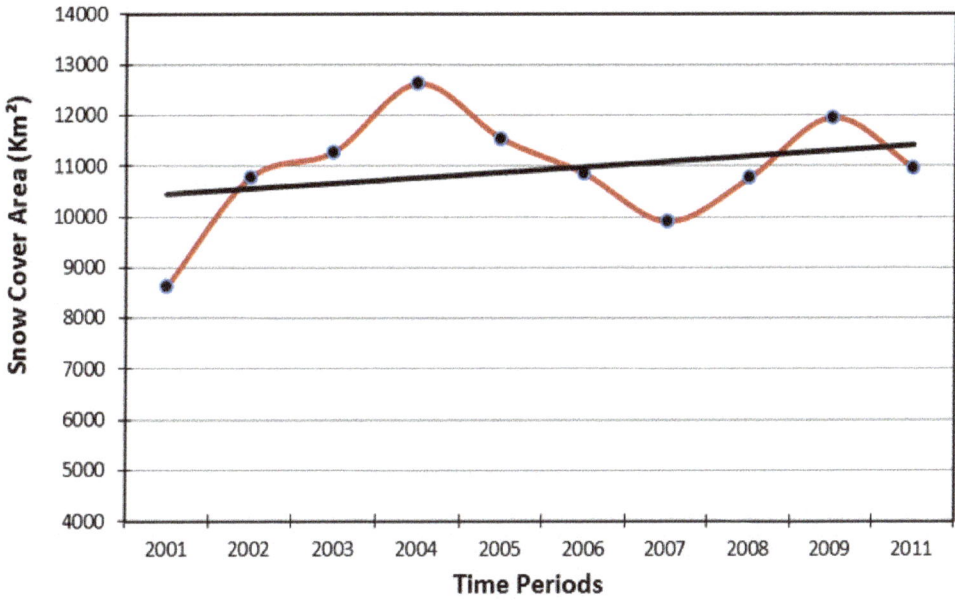

Figure 10. Trend of maximum snow-cover area (SCA) in Hunza basin during the 2001–2011 period.

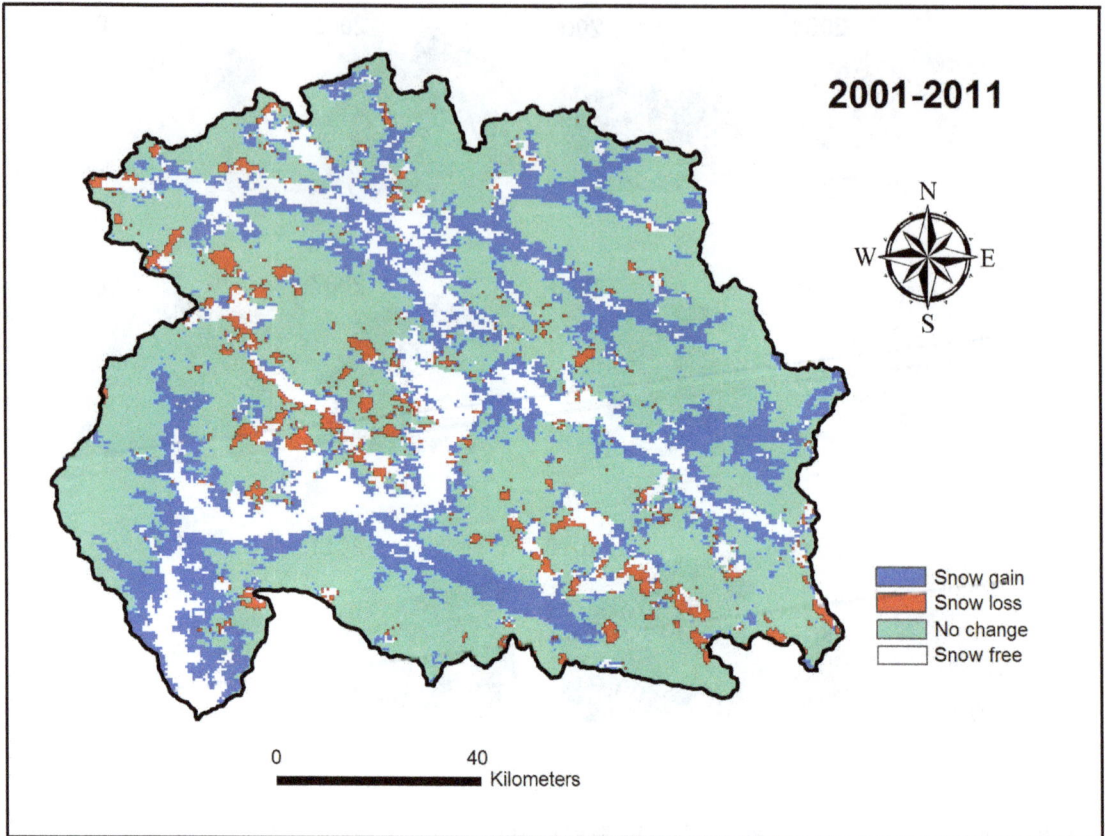

Figure 11. Snow-cover area change (SCAC) during 2001–2011 in Hunza basin.

The snow accumulation has an increasing tendency in the central Karakoram experiencing unique climate signatures, characterized by low temperatures, and enhanced precipitation [52]. The fact also points toward previous observations, e.g., Ref. [53], indicating a regional deviation of the Karakoram glaciers from the usual glacier thinning observed in most glacierized areas of the world and a retreat of some other neighboring Asian glaciers [54]. Since the early 1960s, a rise in winter precipitation in the Karakoram has been observed [55]. From the analysis of interannual variations in the snow-cover area, it has been observed that snow gain is predominantly increasing at the rate of 360 km²/year in this region perhaps because of high elevation and complex orographic features. During the periods of maximum snow gain, snow was found even at the lowest elevation of 1,400 m. Retention of snow at this low-elevation zone indicates heavy snowfall during this period. Snow gain may be characterized by many factors because the high-altitude region is influenced by complex weather systems. From the analysis of snow loss in terms of elevation (extracted from DEM), it was seen that maximum loss occurred at an elevation ranging from 2,000 to 4,000 m during 2002, 2004, 2005, and 2009 probably because of more liquid precipitation than solid within this elevation range. Most of the snow loss was observed within lower elevations of the valley glaciers, such as Batura, Hispar, and Khinyang.

4.2. Analysis of the glacial environment

The identification of glacial features was performed effectively through variable stretching of the pixel values of Landsat ETM+ panchromatic image data (Figure 12). The land features such as ridgelines and drainage network are highlighted in 0–150 stretch in values of panchromatic image. Similarly, the moraine boundaries are distinct in this stretch range providing good approximation of limits of debris-covered glaciers. Low stretch in values (i.e., 0–100) has proved helpful in extracting glacial ice appearance within the high mountainous shadows. Table 2 indicates stretching values of panchromatic image suitable for identification and delineation of various glacial features. The delineation of snow-/ice-covered ridge and catchment boundaries in panchromatic image is possible using greater than 200-value range. The surface variations and flow pattern of glacial ice become highly distinct using this range in the image.

In the Hunza basin, a total of 1,050 glaciers were identified, which contain 10 glaciers of more than 100 km² area (Figure 13), Batura, Hispar, and Hasanabad being the renowned ones. These and some other glaciers in this basin penetrate well below 3,000 m, e.g., the 59-km-long Batura glacier, one of the eight largest glaciers of the middle and low latitudes, has its terminus at about 2,400 m. About two-thirds of the middle and lower parts of the glacier is covered with debris (shown in reddish brown color resembling the surrounding rocks in Figure 14) except for a thin strip of white ice (visible in variable shades of cyan color) that extends to within about 4 km of the terminus. There are seven glaciers that have an area ranging within 50–100 km², whereas 12 glaciers belong to 20–50 km² category. The medium-sized glaciers (10–20 km²) are 24 in number, whereas the rest belong to small-sized glaciers (less than 10 km² in size). According to Hewitt [53], the central Karakoram region is one of the exceptional and exclusive cases throughout the world where the expansion of glaciers has been observed. The Karakoram glaciers are the largest store of moisture in Central Asia and the single-most concentrated source of runoff for the whole Upper Indus basin. The glaciers residing on the steep mountains as well as lying in valleys are highly susceptible to global warming, which may create future hazards for downstream communities (Figure 15a). One of the glaciers of surging type in the Upper Hunza valley, e.g., Ghulkin, has burst several times in the past (recently in early 2015) resulting in a loss of infrastructure, property, and valuable lives (Figure 15b). The glaciers prone to surging or that display irregular flow might be expected to be possible candidates for the generation of outburst floods [13].

In the Astore basin, there were 588 glaciers identified, among which about 99% glaciers belong to 0–10 km² size category, while only 0.2% glaciers belong to 50–100 km² category (Figure 13). No glacier of greater than 100 km² area was found in this basin. The presence of relatively higher numbers of medium- to large-sized glaciers in the Karakoram basin provides an evidence of favorable climate conditions for the glaciers' existence at higher altitudes. The minimum numbers of large-sized glaciers identified in the Himalayan basin point toward higher rates of glacial-ice melting due to increased warming conditions in this range.

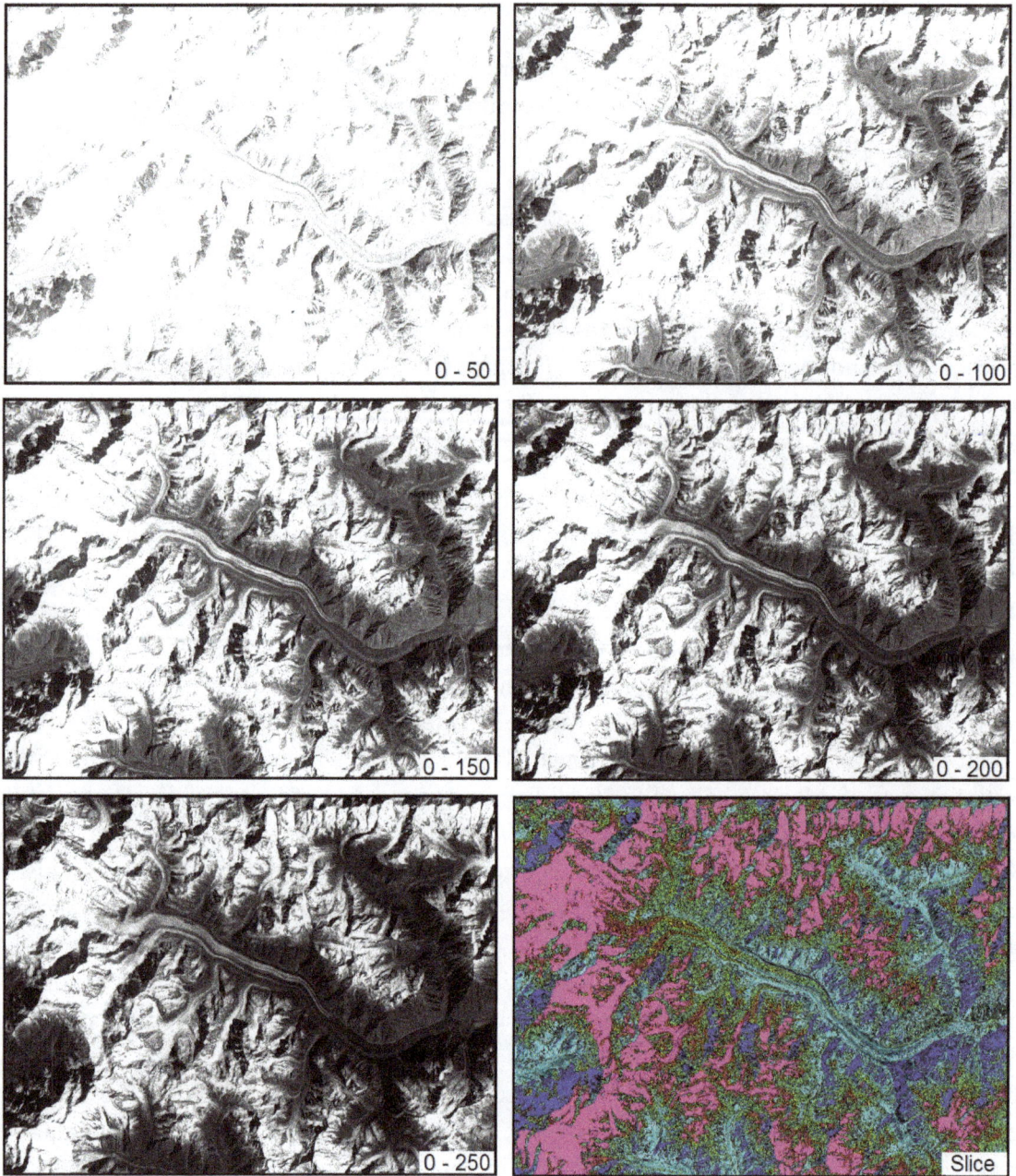

Figure 12. The extraction of glacial features is facilitated by stretching of gray scale values of Landsat ETM+ panchromatic image.

Three basins of the Himalayas, e.g., Shingo, Astore, and Jhelum, were selected to analyze variations in the glacial lakes during 2001 and 2013. Overall, 463 glacial lakes common during the two periods were selected for the analysis. The 204 glacial lakes in Shingo basin indicated an increase in area from 10.35 to 10.84 km^2 (Table 3). The 93 lakes in Astore and 166 lakes in Jhelum basin indicated a minor decrease in coverage during the 12-year period. Overall changes in the lakes area were positive in the three river basins indicating a net expansion in lakes area in the Himalaya range. Variable changes in the lakes area in the basins during the 2001–2013 period are shown graphically in Figure 16 and geographically in Figures 17a–c. The formation of several new glacial lakes is mainly a result of glacier retreat that is observed in most of the Hindu Kush–Karakoram–Himalayan region [19]. The influence of climate on glacial lakes is rather complex and cannot solely account for lake variations [4].

S.No.	Feature	50	100	150	200	250	Slice
1	Exposed ridgeline	L	M	H	M	L	L
2	Ridgeline covered under snow/ice	N	P	L	M	H	N
3	Snow/ice in shadow cover	H	M	L	P	N	P
4	Cascading glacier/ice flow pattern	N	P	L	M	H	P
5	Glacial lake	H	M	L	P	P	L
6	Drainage network	P	M	H	M	L	P
7	Moraine boundary	P	M	H	M	L	P
8	Cloud cover/shadow	M	M	L	L	P	P

H = High; M = Medium; L = Low; P = Poor, N = Nil

Table 2. Suitability of stretching values of panchromatic image for identification of various glacial features.

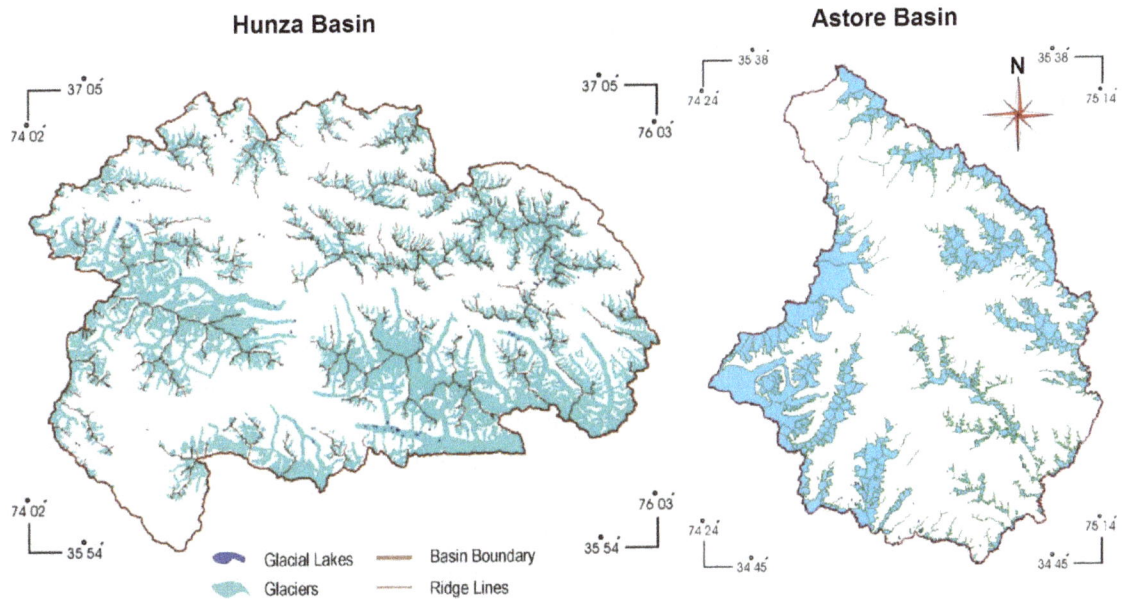

Figure 13. Glaciers and lakes distribution in Hunza and Astore basins.

Figure 14. Landsat ETM+ image of a large valley glacier – Batura in upper Hunza valley.

Basin	Number	Area 2001 (km²)	Area 2013 (km²)	Change (km²)
Shingo	204	10.35	10.84	0.49
Astore	93	4.20	3.98	−0.22
Jhelum	166	10.78	10.52	−0.25
Total	463	25.32	25.34	0.02

Table 3. The glacial lakes status in the Himalayas during 2001 and 2013.

In terms of altitude, the expansion in the lakes area of Shingo basin was positive within the 3,500–5,000 elevation range (Table 4). The expansion in the glacial lakes area within 4,500–5,000 m points toward changes in the climatic pattern, e.g., increase in warming condition resulting in melting of snow/ice or liquid precipitation that might contribute to growth of lakes area. In Astore basin, the change in glacial lakes area was positive within 3,000–4,000 and 4,500–5,000 m elevation ranges (Table 4). The maximum number of glacial lakes within 4,000–4,500 m indicated a decline in area due to the effect of glacial hydrodynamics and/or climatic variations at this elevation range. In Jhelum basin, the change in glacial lakes area was positive within 3,000–3,500 m, while it was negative within 3,500–4,500 m elevation range (Table 4). The maximum number of glacial lakes lie within 4,000–4,500 m in this basin (similar to Astore basin), which also indicated a decline in coverage during the 12-year period.

(a)

(b)

Figure 15. (a): The glaciers descending from steep gradients of Karakoram mountains are susceptible to global warming. (b): Frequency of glacial floods has been increased in the HKH region.

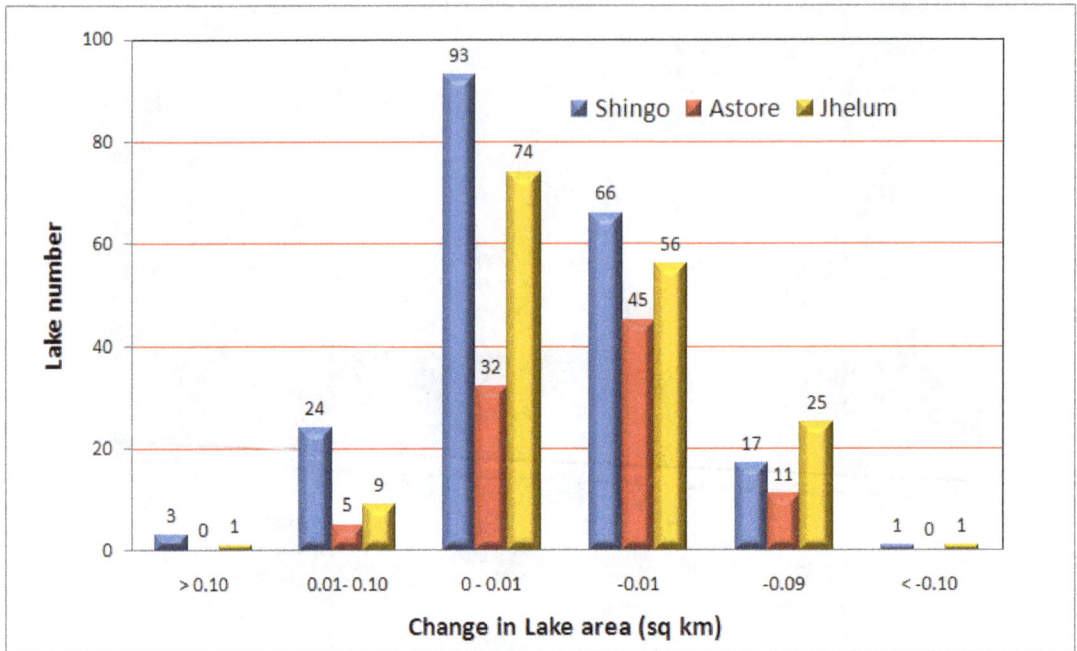

Figure 16. Variable changes in the lakes area in three river basins during the 2001–2013 period.

Elevation (m)	No. of Lakes	Area_2001 (km²)	Area 2013 (km²)	Difference
		Shingo Basin		
3,500–4,000	1	0.01	0.23	0.22
4,000–4,500	75	4.02	4.16	0.14
4,500–5,000	128	6.31	6.45	0.14
Total	**204**			
		Astore Basin		
3,000–3,500	2	0.18	0.21	0.03
3,500–4,000	12	1.37	1.42	0.05
4,000–4,500	62	2.24	1.93	−0.31
4,500–5,000	17	0.41	0.42	0.01
Total	**93**			
		Jhelum Basin		
3,000–3,500	3	1.29	1.46	0.17
3,500–4,000	28	3.27	3.13	−0.14
4,000–4,500	135	6.21	5.93	−0.28
Total	**166**			

Table 4. Changes in the lakes area by elevation in three river basins during the 2001–2013 period

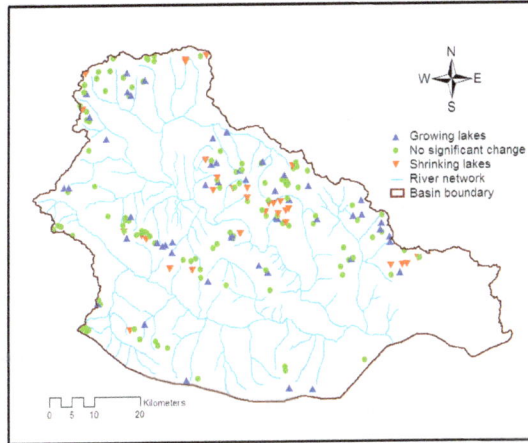

(a)

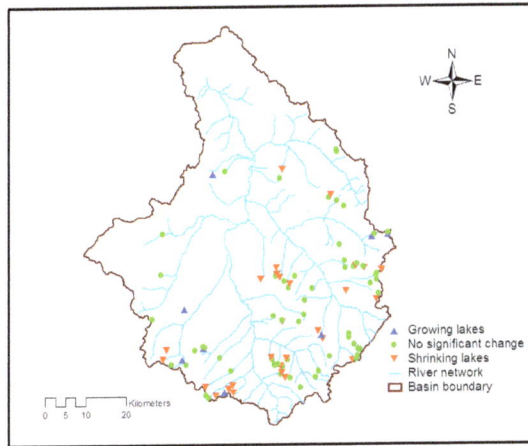

(b)

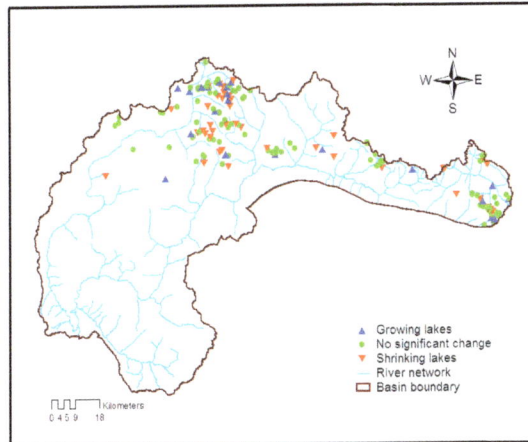

(c)

Figure 17. (a): Changes in the lakes area in Shingo basin during the 2001–2013 period. (b): Changes in the lakes area in Astore basin during the 2001–2013 period. (c): Changes in the lakes area in Jhelum basin during the 2001–2013 period.

The glacier retreat in the Himalayas has resulted in the formation of new glacial lakes and the enlargement of existing ones due to the accumulation of meltwater behind loosely consolidated end-moraine dams [56]. There was a rising trend observed in Astore River flow during the period 1974–2005. The situation may be attributed to the increase in contribution of snow and ice melts in the river flows. The increase in summer temperatures had affected the overall glacial coverage, thickness, and ice reserves during the period 1964–2005. There was a gradual decline in the glacial coverage since 1960s (Figure 18). The melting rates of small glaciers appeared higher than those of the large ones. The trend in depletion of the glacial coverage during 1964–2005 is shown in Figure 19. There are previous studies that highlight the receding of glaciers in most of the Himalaya and a general shrinkage on a global scale [30, 31].

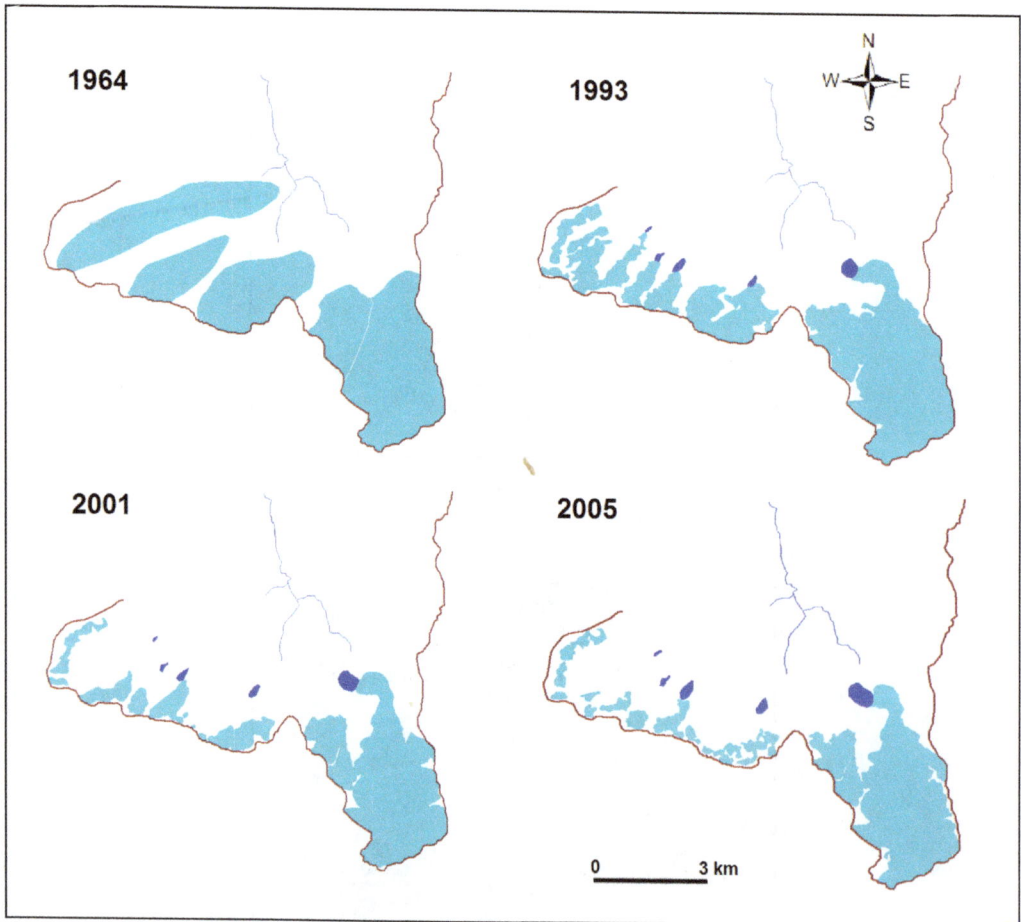

Figure 18. Spatiotemporal analysis of glaciers and glacial lakes in Astore basin.

The melting of glaciers results not only in reduction of surface area and thickness of glaciers but also in expansion of the associated glacial lakes. A large valley glacier *Folvi* (Gr 1) is expanding at a rate of about 0.013 km² y⁻¹ [57], while the depletion of this glacier resulted in expansion of its associated glacial lake at a rate of about 0.009 km² y⁻¹ since 1993. Local geomorphic and climatic parameters may influence the retreat of individual glaciers and may not represent the regional changes in climatic condition.

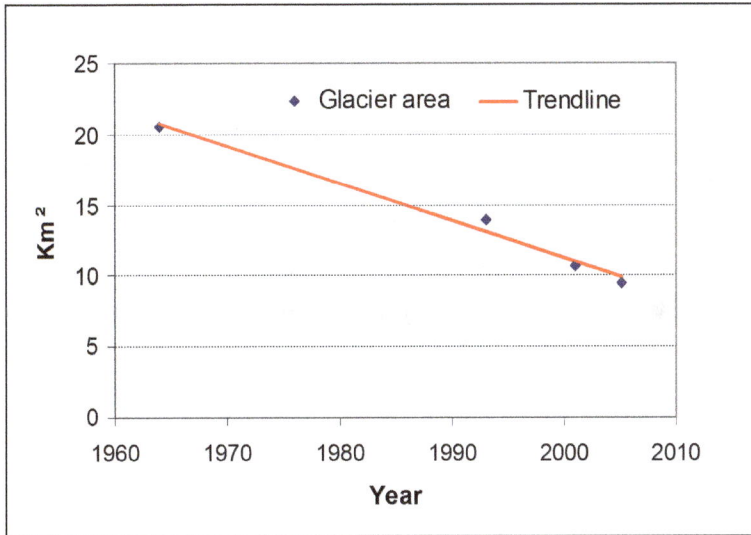

Figure 19. Variation in glacierized area at different time periods in Astore basin.

4.3. Scenarios of impact of climate change

The impact of rise in annual temperature 'T' on river flows was analyzed. It was observed that there was an increasing trend in winter maximum and minimum temperatures of Gilgit from 2011 to 2099. The average values of winter maximum and minimum temperatures were used in snowmelt runoff model to predict future flows of Gilgit River. By increasing the annual maximum and minimum temperatures to 1.24°C until 2050, the summer flows will increase by 16%, and when this temperature increases to 2.78°C untill 2099, summer flows will increase by 34% (Figures 20a&b). Global warming may intensify the summer monsoon, and thereby enhances precipitation especially downstream of the Indus River [25]. The rise in temperature may accelerate the process of seasonal snow and glacier melting resulting in a gradual increase in the river flows.

The results obtained from regional climate model (PRECIS) show that there is an increasing trend of winter precipitation in Gilgit River basin. On this basis, scenarios were developed such that if cryosphere area in Gilgit increases in future due to increase in winter precipitation then what will be its effect on future flows? In previous studies, it was assumed that cryosphere area would increase due to increase in precipitation in Karakorum region as explained in Refs. [53, 58]. Therefore, on this basis, the scenario of 10% increase in the cryosphere area until 2050 and 20% increase in the cryosphere area until 2075 was used in SRM to predict future flows of Gilgit River. According to the modeling results, if cryosphere area increases to 10% and 20% in the basin, summer flows will increase to 13% and 27%, respectively (Figures 20c&d).

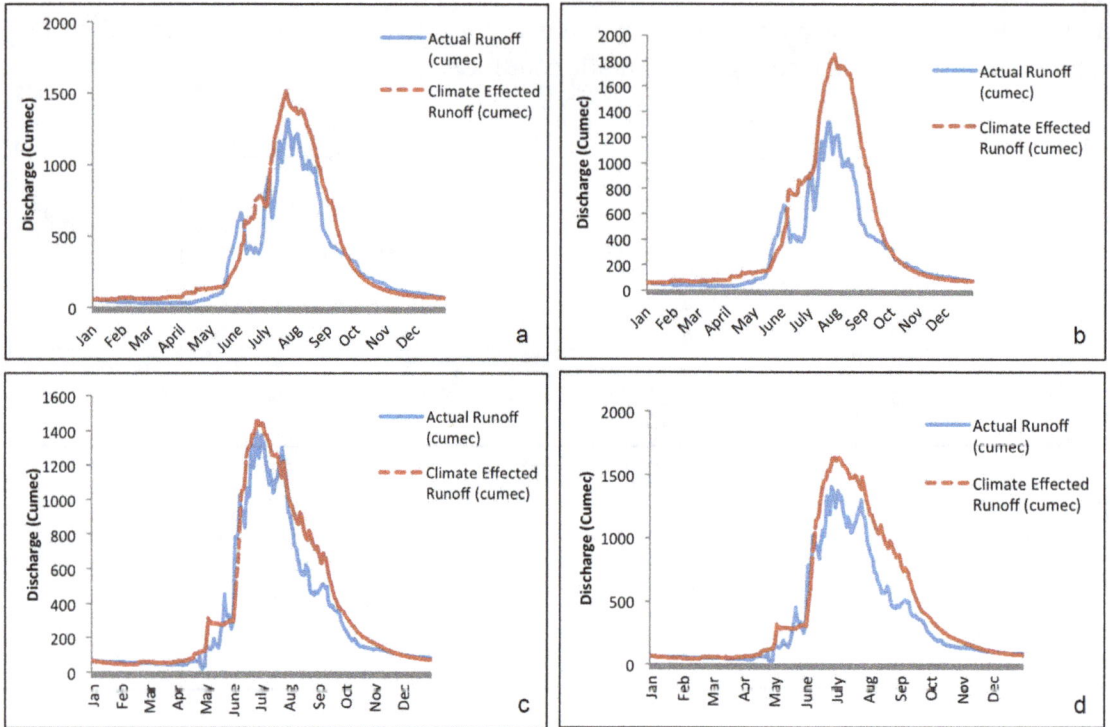

Figure 20. Climate effected runoff due to rise in average annual temperature until 2050 (a); until 2099 (b); increase in cryosphere area by 10% (c); and by 20% (d) in Gilgit basin, Central Karakoram.

4.4. Hazard assessment and early warning using RS technique

The primary functions of remote sensing approach within the climate risk cycle in glacial environment are to develop understanding of the hazards and monitoring the status of glacial lakes and associated glaciers. Predicting whether a glacier will block a valley and, if so, whether a hazardous lake will develop is difficult and requires monitoring of glaciers through physical approach or remote sensing technique. Current hazard assessment efforts depend on detecting margin fluctuations of those glaciers in physiographical settings favorable to lake development observed through field inspection or derived from optical satellite images. Small supraglacial lakes in majority of cases are not hazardous, but they may generate surprisingly large floods that represent hazards at local scales. They can be particularly difficult to identify and assess using remote sensing because of their frequent small size and short life span. Because of the tendency for repeat events from a single glacier, historical reports of GLOFs and local knowledge are important sources of information.

Early warning systems are helpful in reducing the threat of glacial hazards posed to people in downstream. A complete and effective early warning system mainly comprises interrelated elements: risk knowledge, monitoring and warning service, dissemination and communication, and response capability. The human dimensions of early warnings imply that traditional systems are more likely to factor in attachment to the home environment, assets, belief systems, and traditional coping strategies [59]. Although community's involvement in early warning systems is important, remote sensing technology can facilitate in solving the scientific issues related to hazard monitoring, forecasting, and telecommunications.

5. Conclusions

The results of the study reveal that glaciers in this part of Himalayan region are being affected by global warming. Increase in the number of glacial lakes in the recent decade provides clue to the changing glacial environment of the Upper Indus basin. The integrated use of RS and GIS techniques with sparse in situ data is found helpful in analyzing the glaciers' behavior of the Himalayan region. Minimum numbers of large-sized glaciers were identified in the Himalaya basin, which points toward higher rates of glacial ice melting in this range. On the contrary, the presence of relatively higher numbers of medium- to large-sized glaciers in the Karakoram basin provides an evidence of favorable climatic conditions for the glaciers' existence at higher altitudes. Similarly. the increase in snow coverage observed in the Hunza basin of Karakoram during the 2001–2011 may result in ice mass gain in the basin. In order to detect potentially critical glacial lakes in advance, adoption of reliable and robust RS-based approaches is required. The rapidly expanding glacial lakes especially near the headwaters and settlements in the glacierized basins needs to be monitored periodically on a long-term basis to mitigate the risk of any future flood hazards in the HKH region. An in-depth study of the impact of global warming on cryosphere of the Himalayan region using high-resolution remote sensing data (IKONOS, QuickBird, aerial photographs) combined with detailed field investigations is required to cope up with situations such as diminishing water resources and flood hazards in the downstream areas in future.

Acknowledgements

The authors appreciate the support of Pakistan Meteorological Department (PMD), International Center for Integrated Mountain Development (ICIMOD), Nepal, UNDP-Pakistan and Scientists and staff of Climate Change, Alternate Energy and Water Resources Institute for rendering valuable assistance during execution of this study.

Author details

Arshad Ashraf*, Manshad Rustam, Shaista Ijaz Khan, Muhammad Adnan and Rozina Naz

*Address all correspondence to: mashr22@yahoo.com

Climate Change, Alternate Energy and Water Resources Institute (CAEWRI), National Agricultural Research Center, Islamabad, Pakistan

References

[1] Gardelle J, Arnaud Y, Berthier E. Contrasted evolution of glacial lakes along the Hindu Kush Himalaya mountain range between 1990 and 2009. Global and Planetary Change. 2011; 75: 47–55. doi:10.1016/j.gloplacha.2010.10.003

[2] Costa JE and Schuster RL. The formation and failure of natural dams. Geological Society of American Bulletin. 1988; 100: 1054–1068.

[3] Ulugtekin N, Balcik FB, Dogru AO, Goksel C, Alaton IA, Orhon D. The use of remote sensing and geographic information systems for the evaluation of river basins: a case study for Turkey, Marmara River Basin and Istanbul. Journal of Environmental Science and Health, Part A. 2009; 44(4): 388–396. doi:10.1080/10934520802659760

[4] Huggel C, Kääb A, Haeberli W, Teysseire P, Paul F. Remote sensing based assessment of hazards from glacier lake outbursts: a case study in the Swiss Alps. Canadian Geotechnical Journal. 2002; 39: 316–330.

[5] Jacobs JD, Simms EL, Simms A. Recession of the southern part of Barnes Ice Cap, Baffin Island, Canada, between 1961 and 1993, determined from digital mapping of Landsat TM. Journal of Glaciology. 1997; 43: 98–102.

[6] Kääb A, Paul F, Maisch M, Hoelzle M, Haeberli W. The new remote-sensing-derived Swiss glacier inventory: II. First results. Annals of Glaciology. 2002; 34: 362–366.

[7] Roohi R, Mool PK, Ashraf A, Bajracharya S, Hussain SA, Naz R, 2005. Inventory of Glaciers, Glacial lakes the Identification of Potential Glacial lake Outburst Floods Affected by Global Warming in the Mountains of Himalayan Region, Pakistan, ICIMOD, Nepal and PARC, Pakistan.

[8] Williams RS Jr. and Ferrigno JG (eds.), State of the Earth's cryosphere at the beginning of the 21st century – glaciers, global snow cover, floating ice, and permafrost and periglacial environments: U.S. Geological Survey Professional Paper 1386–A: 546 p. 2012. http://pubs.usgs.gov/pp/p1386a.

[9] Quincey D, Richardson S, Luckman A, Lucas R, Reynolds J, Hambrey M, Glasser N. Early recognition of glacial lake hazards in the Himalaya using remote sensing datasets. Global and Planetary Change. 2007; 56: 137–152.

[10] Kääb A. Glacier volume changes using ASTER satellite stereo and ICESat GLAS laser altimetry. A test study on Edgeøya, Eastern Svalbard. IEEE Transactions on Geoscience and Remote Sensing. 2008; 46(10): 2823–2830.

[11] Veettil BK. A Remote sensing approach for monitoring debris-covered glaciers in the high altitude Karakoram Himalayas. International Journal of Geomatics and Geosciences. 2012; 2(3): 833–841.

[12] Richardson SD. Remote sensing approaches for early warning of GLOF hazards in the Hindu Kush-Himalayan region, Final report-ver 1.2, United Nations International Strategy for Disaster Reduction (UN/ISDR). 2010.

[13] Huggel C, Kääb A, Haeberli W, Krummenacher B. Regional-scale GIS-models for assessment of hazards from glacier lake outbursts: evaluation and application in the Swiss Alps. Natural Hazards and Earth System Sciences. 2003; 3: 647–662.

[14] Martinec J. Snowmelt runoff model for stream flow forecasts. Nordic Hydrology. 1975; 6(3): 145–154. doi:10.2166/nh.1975.010

[15] Poon SKM. Hydrological Modeling Using MODIS Data for Snow Covered Area in the Northern Boreal Forest of Manitoba. M.E. Dissertation. Alberta: University of Calgary. 2004.

[16] Klein AG and Barnett AC. Validation of daily MODIS snow cover maps of the Upper Rio Grande River Basin for the 2000-2001 snow year. Remote Sensing of Environment. 2003; 86(2): 162–176. doi:10.1016/ S0034-4257(03)00097-X

[17] Haeberli W. Changing views on changing glaciers. In The Darkening Peaks: Glacial Retreat in Scientific and Social Context (eds.) Orlove B, Wiegandt E and Luckman B, Berkeley: University of California Press, 2005: 23–32.

[18] Haeberli W and Beniston M. Climate change and its impacts on glaciers and permafrost in the Alps. Ambio. 1998; 27: 258–265.

[19] GCOS: Implementation plan for the Global Observing System for Climate in support of the UNFCCC, GCOS–92, WMO, Geneva. 2004.

[20] IPCC. Fifth Assessment Report of Intergovernmental Panel on Climate Change: Working Group 1. Summary for Policy Makers – Climate Change 2013: The Physical Science Basis. 2013.

[21] UNEP and WGMS. Global glacier changes: facts and figures. 2008.

[22] Jansson P, Hock R, Schneider T. The concept of glacier storage: a review. Journal of Hydrology. 2003; 282(1–4): 116–129.

[23] Sharif M, Archer D, Fowler H, Forsythe N. Trends in timing and magnitude of flow in the Upper Indus Basin. Hydrology and earth system sciences discussions. 2012; 9: 9931–9966.

[24] Bocchiola D, Diolaiuti G, Soncini A, Mihalcea C, Agata CD, Mayer C, Lambrecht A, Rosso R, Smiraglia C. Prediction of future hydrological regimes in poorly gauged high altitude basins: the case study of the upper Indus, Pakistan. Hydrology and Earth System Sciences Discussions. 2011; 8(2): 3743–791.

[25] Panday P, Frey K, Ghimire B. Detection of the timing and duration of snowmelt in the Hindu Kush-Himalaya using QuikSCAT, 2000–2008. Environmental Research Letters. 2011; 8(1): 014020–014033.

[26] Frey H. Compilation and Applications of Glacier Inventories using Satellite Data and Digital Terrain Information, PhD Dissertation University of Zurich. 2011.

[27] Posma JC. The effects of climate change and associated glacier melting in the Hindu-Kush Himalayas on the water supply and water use of the Indus, Pakistan. BS Thesis, University of Utrecht. 2013.

[28] Ashraf A, Roohi R, Naz R, Mustafa N. Monitoring cryosphere and associated flood hazards in high mountain ranges of Pakistan using remote sensing technique. Natural Hazards. 2014; 73: 933–949. doi:10.1007/s11069-014-1126-3

[29] Chaudhry QZ, Mahmood A, Rasul G, Afzaal M. Climate Indicators of Pakistan. PMD Technical Report 22/2009.

[30] Fujita K and Nuimura T. Spatially heterogeneous wastage of Himalayan glaciers. Proceedings of the National Academy of Sciences of the United States of America. 2011; 108(34): 14011–14014. doi:10.1073/pnas.1106242108

[31] Bolch T, Kulkarni A, Kaab A, Huggel C, Paul F, Cogley JG, Frey H, Kargel JS, Fujita K, Scheel M, Bajracharya S, Stoffel M. The state and fate of Himalayan glaciers. Science. 2012. 336(6079): 310–314.

[32] Kääb A, Berthier E, Nuth C, Gardelle J, Arnaud Y. Contrasting patterns of early twenty-first-century glacier mass change in the Himalayas. Nature. 2012; 488: 495–498. doi:10.1038/nature11324

[33] Gardelle J, Berthier E, Arnaud Y, Kääb A. Region-wide glacier mass balances over the Pamir-Karakoram-Himalaya during 1999–2011. The Cryosphere. 2013; 7: 1263–1286. doi:10.5194/tc-7-1263-2013

[34] WMO. Intercomparison of Models of Snowmelt Runoff, Geneva, Switzerland. 1986.

[35] Hewitt K. Tributary glacier surges: an exceptional concentration at Panmah glacier Karakoram Himalaya. Journal of Glaciology. 2007; 53: 181–188.

[36] Archer DR. The climate and hydrology of northern Pakistan with respect to assessment of flood risks to hydropower schemes. Report by GTZ/WAPDA. 2001.

[37] Awan SA. The climate and flood risk potential of northern Pakistan, special issue of Journal of Science vision. 2002; 47(3,4): 100–109.

[38] LIGG/WECS/NEA. Report on the First Expedition to Glaciers and Glacier Lakes in the Pumqu (Arun) and Poique (Bhote-Sun Kosi) River Basins, Xizang (Tibet), China, Sino-Nepalese Investigation of Glacier Lake Outburst Floods in the Himalaya. Beijing, China: Science Press. 1988.

[39] Wang WC, Yao TD, Yang XX. Variations of glacial lakes and glaciers in the Boshula mountain range, southeast Tibet, from the 1970s to 2009. Annals of Glaciology. 2011; 52: 9–17. doi:10.3189/172756411797252347

[40] Salerno F, Thakuri S, D'Agata C, Smiraglia C, Manfredi EC, Viviano G, Tartari G. Glacial lake distribution in the Mount Everest region: uncertainty of measurement and conditions of formation. Global and Planetary Change. 2012; (92–93): 30–39. doi: 10.1016/j.gloplacha.2012.04.001

[41] Pietroniro A and Leconte R. A review of Canadian remote sensing applications in hydrology, 1995–1999. Hydrological Processes. 2000; 14: 1641–1666..

[42] Bayr KJ, Hall DK, Kovalick WM. Observations on glaciers in the eastern Austrian Alps using satellite data. International Journal of Remote Sensing. 1994; 15(9): 1733–1752.

[43] Paul F. Evaluation of different methods for glacier mapping using Landsat TM. EARSeL eProc. 2001; 1: 239–245.

[44] Paul F, Kääb A, Maisch M, Kellenberger T, Haeberli W. The new remote-sensing-derived Swiss glacier inventory. I. Methods. Annals of Glaciology. 2002; 34: 355–361.

[45] Crane RB. Preprocessing techniques to reduce atmospheric and sensor variability in multispectral scanner data. In Proceedings of the 7th International Symposium on Remote Sensing of the Environment, Vol. II, Ann Arbor, MI, USA. 1971: 1345–1355.

[46] Vincent RK. Spectral ratio imaging methods for geological remote sensing from aircraft and satellites. In Anson A ed. Proceedings of the American Society of Photogrammetry, Management and Utilization of Remote Sensing Data Conference, Sioux Falls, South Dakota. Falls Church, VA, American Society of Photogrammetry. 1973: 377–397.

[47] Holben BN. An examination of spectral band rationing to reduce the topographic effect on remotely sensed data. International Journal of Remote Sensing. 1981; 2(2): 115–133.

[48] Meyer P, Radiometric corrections of topographically induced effects on Landsat TM data in an alpine environment. ISPRS Journal of Photogrammetry. 1993; 48(4): 17–28.

[49] Muller F, Caflish T, Muller G. Instruction for Compilation and Assemblage of Data for a World Glacier Inventory. Zurich: Temporary Technical Secretariat for World Glacier Inventory, Swiss Federal Institute of Technology, Zurich. 1977.

[50] Tahir AA, Chevallier P, Arnaud Y, Neppel L, Ahmad B. Modeling Snowmelt-Runoff under climate scenarios in the Hunza River basin, Karakoram Range, Northern Pakistan. Journal of Hydrology. 2011; 409(1–2): 104–117.

[51] Farooqi AB, Khan AH, Mir H. Climate change perspective in Pakistan. Pakistan Journal of Meteorology. 2005; 2(3): 11–21.

[52] Fowler HJ and Archer DR. Conflicting signals of climate change in the upper Indus basin. Journal of Climate. 2005; 9: 4276–4293.

[53] Hewitt K. The Karakoram anomaly? Glacier expansion and the 'Elevation Effect', Karakoram Himalaya. Mountain Research and Development. 2005; 25: 332–340.

[54] IPCC. Climate Change 2007: Contribution of Working Groups I, II and III to the Fourth Assessment Report of the Intergovernmental Panel on Climate Change. Core Writing Team (eds.) Pachauri RK and Reisinger A, Geneva: IPCC, 2007: pp 104.

[55] Archer DR and Fowler HJ. Spatial and temporal variations in precipitation in the Upper 20 Indus Basin, global teleconnections and hydrological implications, Hydrology and Earth System Sciences. 2004; 8: 47–61.

[56] ICIMOD. Glacial lakes and glacial lake outburst floods in Nepal. Kathmandu: ICIMOD. 2011.

[57] Ashraf A, Roohi R, Ijaz S, Ahmad B, Naz R. Monitoring Global warming Impact on Glacier environment using GIS Application. In proceedings of National Seminar on State and Challenges of GIS/RS Applications in Water sector, by PCRWR, Islamabad 25–26 June, 2008.

[58] Immerzeel WW, Van Beek LPH, Bierkens MFP. Climate change will affect the Asian water towers. Science. 2010; 328(5984): 1382–1385.

[59] ISDR. Living with Risk. A Global Review of Disaster Reduction Initiatives. Inter-Agency Secretariat of the International Strategy for Disaster Reduction (ISDR). Geneva, 2007: pp 384.

4

Topological Characterization and Advanced Noise-Filtering Techniques for Phase Unwrapping of Interferometric Data Stacks

Pasquale Imperatore and Antonio Pepe

Additional information is available at the end of the chapter

Abstract

This chapter addresses the problem of phase unwrapping interferometric data stacks, obtained by multiple SAR acquisitions over the same area on the ground, with a twofold objective. First, a rigorous gradient-based formulation for the multichannel phase unwrapping (MCh-PhU) problem is systematically established, thus capturing the intrinsic topological character of the problem. The presented mathematical formulation is consistent with the theoretical foundation of the *discrete calculus*. Then within the considered theoretical framework, we formally describe an innovative procedure for the noise filtering of time-redundant multichannel multilook interferograms. The strategy underlying the adopted multichannel noise filtering (MCh-NF) procedure arises from the key observation that multilook interferograms are not fully time consistent due to multilook operations independently applied on each single interferogram. Accordingly, the presented MCh-NF procedure suitably exploits the temporal mutual relationships of the interferograms. Finally, we present some experimental results on real data and show the effectiveness of our approach applied within the well-known small baseline subset (SBAS) processing chain, thus finally retrieving the relevant Earth's surface deformation time series for geospatial phenomena analysis and understanding.

Keywords: SAR interferometry, phase unwrapping, discrete calculus

1. Introduction

Multichannel (or multitemporal) InSAR techniques address the processing of interferometric data stacks obtained by combining multiple SAR acquisitions over the same area. These approaches can be essentially categorized in two main classes: persistent scatterers (PS) and small baseline (SB)-based techniques. The solution of the multichannel phase unwrapping (MCh-PhU) problem is generally required in multichannel InSAR techniques, whenever multidimensional SAR data set, conveying information about complex Earth's crust events, have to be systematically investigated on suitable space-time scales for geospatial phenomena understanding [1–29]. In this chapter, we focus on two different related main issues.

Primarily, we present a rigorous gradient-based formulation of the MCh-PhU problem that is consistent with the theoretical foundation of the discrete calculus [30–34]. Emphasis is placed on the topological characterization of the underlying discrete setting provided by the *differential operators* of the discrete calculus, which are formally associated with matrix operators and represent the discrete counterparts of the classical differential operators of the vector calculus. Accordingly, MCh-PhU problem formulation is systematically established in terms of discrete differential operators, which are defined by the topology of the intrinsically discrete spaces upon which they act, thus capturing the essential topological character of the problem within a systematic matrix formalism [35]. It is worth highlighting that our approach provides an unambiguous and theoretical-consistent formalism for the MCh-PhU problem, overcoming the conceptual inconsistencies of the existing gradient-based formulations [1, 17, 29]. Indeed, the existing approaches pose some conceptual limitations from a mathematical viewpoint since they rely on an intrinsically discrete setting based description and, at the same time, resort to the concepts of gradient and curl of the conventional vectorial calculus, which inherently imply a reference to an underlying continuum space and the notion of the infinitesimal [30]. In addition, the proposed formal framework enables meaningful analytical investigations on a mathematical consistent playground, also providing interesting implications and permitting to express previous obtained results in a more general way.

Then we present an innovative procedure to filter out the noise affecting the phase components of a redundant set of (multitemporal) multilook small-baseline interferograms. This is achieved by independently solving, for each pixel of the scene, a *nonlinear optimization* problem based on computing the wrapped phase vector that minimizes the (weighted) circular variance of the difference between the original and noise-filtered interferograms [43]. This noise-filtering procedure arises from the key observation that multilook interferograms are not fully time consistent because they are generated through multilook operations that are independently carried out on each single interferogram. Indeed, the wrapped discrete curl of the interferometric phases defined on a graph whose nodes and edges describe SAR acquisitions (in the time/perpendicular baseline domain) and inherent interferograms, respectively, is different from zero. This modulo-2π cyclic inconsistency of multichannel interferometric phases is properly handled by the presented multichannel noise-filtering (MCh-NF) procedure. The presented technique is very easy to implement because it does not imply a preliminary time-consuming selection of statistically homogenous pixels (SHP), as for instance required by the *SqueeSAR* technique [44], and it has no need of any *a priori* information on the statistics of complex-valued SAR images. The effectiveness of the presented noise-filtering approach as well as its impact on the quality of multichannel phase unwrapping procedures are also fully investigated.

2. Multichannel phase unwrapping problem

In this Section, we review the mathematical formulation of the multichannel phase unwrapping (MCh-PhU) problem within the purview of the discrete calculus. As a matter of fact, a graph-based description naturally arises in formulating the MCh-PhU problem due to the underlying discrete irregular data structure. Indeed, as far as discrete settings (e.g., graphs) are concerned, resorting to the conventional vectorial calculus might not be adequate since it inherently implies a reference to an underlying continuum space. On the contrary, *discrete calculus* offers a rigorous methodological framework since it treats a discrete domain as entirely its own entity. In particular, discrete calculus provides proper differential operators that make it possible to purely operate onto a finite, discrete structure without referring to the continuous space and notion of the infinitesimal [30]. More specifically, the introduction of some well-known algebraic structures [30–33] capturing the essential topological character of the underlying graphs permits to phrase pertinent *differential operators* as matrices. Therefore, one of the most important consequences of this approach is that the purely topological nature of the discrete differential operators is made more apparent and concrete. Accordingly, by systematically adopting the relevant key concepts and tools available within this theoretical framework, we here provide a description of the MCh-PhU problem on a suitable mathematical playground. For such a purpose, we first establish the notation and terminology used throughout the subsequent Sections 2.1, and then the problem at hand is reviewed within this formalism (Sections 2.2 and 2.3). We remark that the focus here is on presenting key concepts that are useful for the following analyses; however, a comprehensive treatment of the discrete calculus and related huge fields of mathematics (e.g., algebraic topology, exterior calculus, and differential forms) is clearly beyond the scope of this work but can be found in refs. [30–33].

2.1. The theoretical framework of discrete calculus

A graph $\mathcal{G}(\mathcal{V};\mathcal{E})$ is defined by two sets: $\mathcal{V}$ and $\mathcal{E}$. The former is the set of nodes (or vertices) of the graph, and the latter represents the corresponding set of edges. Let Q and M be the cardinality of $\mathcal{V}$ and $\mathcal{E}$, respectively. The vector space $\mathbb{R}^M$ is referred to as the *edge space*, and the vector space $\mathbb{R}^Q$ is referred to as the *vertex space*, with $\mathbb{R}$ denoting the field of real numbers. Without loss of generality, we here assume that the graph $\mathcal{G}$ is *connected* (i.e., every pair of vertices in the graph is connected [33]). Moreover, an orientation establishes a default direction on an edge that is considered positive or negative, thus yielding an oriented graph. The $M \times Q$ *incidence matrix* $\Pi = [\Pi_{mq}]$ of an oriented graph $\mathcal{G}$ specifies its edge–node connectivity relations, whose entries are defined as follows [30–33]:

$$\Pi_{mq} \equiv \begin{cases} -1, & \text{if } m\text{th edge starts at } q\text{th node} \\ +1, & \text{if } m\text{th edge ends at } q\text{th node} \\ 0, & \text{otherwise} \end{cases} \qquad (1)$$

with $m = 1, 2,..., M$ and $q = 1, 2,..., Q$. It is important to note that the column rank of Π is $Q - 1$. The incidence matrix Π generates an orthogonal decomposition $\mathbb{R}(\Pi) \oplus \mathcal{N}(\Pi^T) = \mathbb{R}^M$, where $\mathbb{R}(\Pi)$ is the column space of the incidence matrix Π, and $\mathcal{N}(\Pi^T)$ denotes the kernel (or null-space) of the matrix $\boldsymbol{\Pi}^T$. The notion of cycle space is also fundamental in graph theory. The *cycle space* of the graph $\mathcal{G}$, namely, $C = \mathcal{C}(\mathcal{G})$, is the subspace of edge space $\mathbb{R}^M$ spanned by all the cycles in $\mathcal{G}$. The dimension of $\mathcal{C}(\mathcal{G})$ is also referred to as the *cyclomatic number* of $\mathcal{G}$ [33]. It is also well known that, for every *connected* graph $\mathcal{G}$ with Q nodes and M edges, the dimension of the cycle space is given by $R = \dim(\mathcal{C}(\mathcal{G})) = M - Q + 1$ [30]. Each basis for $\mathcal{C}(\mathcal{G})$ (i.e., the cycle basis) is therefore uniquely specified by an $M \times R$ matrix $\boldsymbol{\Omega}$, called *cycle matrix*. Thus, the column vectors of $\boldsymbol{\Omega} = [\omega^1, ..., \omega^R]$ form a basis for an R-dimensional vector subspace (the cycle space of $\mathcal{C}(\mathcal{G})$ of $\mathbb{R}^M$. $\mathcal{C}(\mathcal{G})$ is indeed the null-space of $\boldsymbol{\Pi}^T$ so that a cycle basis provides a basis for $\mathcal{N}(\boldsymbol{\Pi}^T)$) [30]. Accordingly, a fundamental property of a linear graph is expressed by the remarkable relations:

$$\boldsymbol{\Pi}^T \boldsymbol{\Omega} = \boldsymbol{0} \tag{2}$$

$$\boldsymbol{\Omega}^T \boldsymbol{\Pi} = \boldsymbol{0} \tag{3}$$

Indeed, several methods for defining a cycle set have been studied, and they can be used to define incidence relations between edges and cycles. Specifically, the definition of a cycle set from the edge set can be obtained algebraically and geometrically (i.e., from an embedding). Algebraic methods find a suitable $M \times R$ matrix $\boldsymbol{\Omega}$ whose columns provide a basis for the null-space of $\boldsymbol{\Pi}^T$, with $R = \dim(\mathcal{N}(\boldsymbol{\Pi}^T))$. Geometric methods for defining a cycle set (i.e., from an embedding) permit to identify algorithmically a cycle set (representing the faces) in this embedding and may be used to produce the edge–face incidence matrix $\boldsymbol{\Omega}$ (as illustrated in Figure 1). In particular, it is possible to consider a normalized irreducible cycle basis forming elementary (or irreducible) cycles [30–33], i.e., cycles that contain no other cycles, so that we can associate to each elementary cycle an elementary cycle vector $\omega^r = [\omega_1, ..., \omega_M]^T$, whose entries are defined as follows:

$$\omega_i \equiv \begin{cases} +1 & \text{if } r\text{th cycle traverses } i\text{th edge forward} \\ -1 & \text{if } r\text{th cycle traverses } i\text{th edge backward} \\ 0 & \text{otherwise} \end{cases} \tag{4}$$

Accordingly, the so defined $\boldsymbol{\Omega} = [\omega^1, ..., \omega^R]$ provides a particularly attractive basis for $\mathcal{N}(\boldsymbol{\Pi}^T)$, i.e., the cycle basis formed by all the elementary cycle vectors associated with the elementary cycles in $\mathcal{G}$. We also note that $\boldsymbol{\Omega}$ defines the incidence connectivity relations between edges and cycles (see Figure 1).

It is instructive to highlight that the topological operators $\boldsymbol{\Pi}$, $\boldsymbol{\Pi}^T$, and $\boldsymbol{\Omega}$ provide the discrete counterparts of the classical *gradient* (∇), *divergence* ($\nabla \cdot$), and *curl* ($\nabla \times$) operators of the vector calculus for continuous space, respectively. Accordingly, they can be regarded as differential operators on the discrete setting [30]. In addition, it is worth emphasizing that identities (2) and (3) mimic the properties of their classical vector calculus counterparts $\nabla \cdot \nabla \times = 0$ (*div curl* = 0) and $\nabla \times \nabla = 0$ (*curl grad* = 0), respectively. It should be pointed out that Π yields differences along edges of nodal "potentials" represented by Πx. Conversely, given an arbitrary $f \in \mathbb{R}^M$, a solution of the equation $\Pi x = f$ (if it exists) is called the *potential* of f. Note also that x (if it exists) is not unique since the constant column $\mathbb{1} = [1, \dots, 1]^T \in \mathbb{R}^Q$ is an element of the kernel of Π. Of course, not every $f \in \mathbb{R}^M$ is the discrete gradient of some x since f may contain a curl component. Indeed, a prescribed $f \in \mathbb{R}^M$ can be written as a nodal difference ($f = \Pi x$) if it is *cyclically consistent*, i.e., if it satisfies $\Omega^T f = 0$ (i.e., there is no component of the flow in the cycle space). Note also that Ω is the (cross) differential operator of the graph whose expression can be given in terms of a normalized cycle basis; $\mathcal{N}(\Omega^T)$ denotes the subspace of $\mathbb{R}^M$ with zero flow circulation (curl-free) around cycles. Moreover, $\Pi^T f$ yields nodal accumulations from flows along edges. As a result, the differential operators, as basic tools of the discrete calculus, have been established and properly phrased on the discrete space. This mathematical abstraction meaningfully captures the topological structure of the underlying discrete setting. Note that the topological characterization of the graph is essentially embodied in the algebraic structure of the associated discrete (matrix) operators and their interrelations. We also stress the significant distinction between the discrete operators and the commonly adopted discretized versions of the continuous differential operations obtained via the method of finite differences in numerical analysis; the latter generally lack the desirable topological behaviour [42].

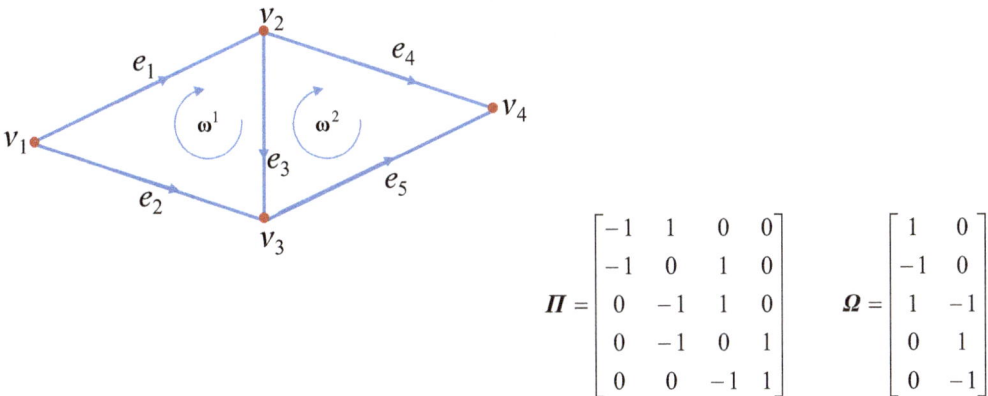

$$\Pi = \begin{bmatrix} -1 & 1 & 0 & 0 \\ -1 & 0 & 1 & 0 \\ 0 & -1 & 1 & 0 \\ 0 & -1 & 0 & 1 \\ 0 & 0 & -1 & 1 \end{bmatrix} \qquad \Omega = \begin{bmatrix} 1 & 0 \\ -1 & 0 \\ 1 & -1 \\ 0 & 1 \\ 0 & -1 \end{bmatrix}$$

Figure 1. An example graph shown along with its edge–node incidence matrix, Π, and cycle (edge-face incidence) matrix, Ω. Note that $M=5$, $Q=4$ and $R=2$.

As a final remark, some considerations on the *total unimodularity* (TU) property inherent to the matrix operators, which is extremely important in integer linear programming, are in order. We recall that a matrix is TU if the determinant of every submatrix is either zero or ± 1. For any graph, the edge–node incidence matrix is TU. On the contrary, the face–edge incidence matrix, in general, is not TU [30–34]. However, the edge–face incidence matrix is TU when each edge is included in exactly two faces that traverse the edge in opposite directions (e.g., a planar graph with a minimum cycle basis [30]). In this circumstance, total unimodularity of the edge–face incidence matrix stems from the fact that the face–edge incidence matrix is the edge–node incidence matrix of the dual graph [30]. Indeed, a TU constraint matrix (and integer constraints) guarantees that the solution of the related optimization problem (see also optimization problems (21) and (22) in the following) will be integer. Nonetheless, TU property has a further practical significance since the relaxed problem, obtained by neglecting the integer constraints, can also be solved using generic linear (not integer) programming solvers.

2.2. Rigorous gradient-based formulation of the MCh-PhU problem

Once the basic concepts of the discrete calculus and graph theory are presented, we are in the position to frame the formulation of the MCh-PhU problem on an appropriate mathematical playground. Let us consider a set of Q *single-look-complex* (SLC) SAR data acquired over a certain area of interest. One of them is assumed as the reference (master) image, with respect to which the images are properly coregistered. This set is characterized by the corresponding acquisition times $\{t_1, \ldots, t_Q\}$ and perpendicular baselines $\{b_{\perp 1}, \ldots, b_{\perp Q}\}$. Accordingly, for each coregistered SLC pair, a multilook *phase interferogram* (suitably depurated by the flat-Earth and topography contributions, by using a priori information about the topography and acquisition geometry) can be produced [37]; however, a common practice (within the multitemporal SB InSAR class) is first to identify a suitable small-baseline subset of the relevant multibaseline (temporal and spatial-perpendicular baselines) interferometric-pair set [6]. This is done to confine the effect of decorrelation phenomena associated with inherent angular and temporal electromagnetic backscattering variations [38]. Furthermore, a subset of common pixels of the M interferograms are then usually identified via the estimated *coherence* [37,38], so that only P pixels characterized by relatively high coherence values are singled out. In other words, the coherence index, providing a quantitative estimation of the decorrelation effects, permits to discriminate in favor of the "reliable" pixels.

The final aim is to reconstruct the absolute (i.e., not restricted in the principal $[-\pi, \pi)$ interval) interferometric phases values from the wrapped (i.e., observed only in the principal $[-\pi, \pi)$ interval) interferometric phase pertinent to M multichannel interferograms.

The problem we are interested in can be naturally framed on a discrete setting. Indeed, one possibility is to regard the discrete set of P selected (typically sparsely distributed) coherent pixels as a set of nodes, $\mathcal{V}_B$, in the Euclidean (*azimuth, range*) plane, and the set of Q SAR acquisitions representing a set of nodes, $\mathcal{V}_A$, in the Euclidean (*temporal-baseline, perpendicular-baseline*) plane [26]. Accordingly, a formal description of the problem at hand can be given in terms of a couple of abstract graphs: $\mathcal{G}_A = (\mathcal{V}_A; \mathcal{E}_A)$ and $\mathcal{G}_B = (\mathcal{V}_B; \mathcal{E}_B)$ where the corresponding edge sets $\mathcal{E}_A$ and $\mathcal{E}_B$ have to be properly defined. Accordingly, with Q and M, we denote the cardinality of $\mathcal{V}_A$ and $\mathcal{E}_A$, and with P and N the cardinality of $\mathcal{V}_B$ and $\mathcal{E}_B$, respectively. Note also that defining $\mathcal{E}_A$ pertains to the M interferometric pairs selection. Defining a meaningful edge set for a collection of nodes is now concerned since different criteria can be adopted to achieve it. The dimensionality of the ambient space in which the graph is embedded deserves some considerations. In this regard, we recall that a graph is called *planar* if it can be embedded in the plane [30–33]. Note also that a graph is not generally guaranteed to be planar, even if the nodes are embedded in two dimensions. Since planarity is important for the workability of the implicated optimization procedure with powerful numerical solvers, a typical option to preserve graph planarity is resorting to the *Delaunay triangulation* in the Euclidean plane for establishing the edge set from nodes embedded in two dimensions [26]. Note that such an option specifically pertains to the solution strategy [26, 35]; nonetheless, our general formulation applies as well when different edge structures are adopted. Accordingly, once $\mathcal{G}_A$ and $\mathcal{G}_B$ have been somehow defined, the topological properties inherent to each graph are algebraically captured by the related differential operators, which are summarized in Table 1.

Symbol	Quantity	Related meaning
Q	Nodes number of $\mathcal{G}_A$	Number of SAR acquisitions
M	Edges Number of $\mathcal{G}_A$	Number of interferometric pairs
R	Dimension of the cycle space of $\mathcal{G}_A$	
Π_A	$M \times Q$ incidence matrix of $\mathcal{G}_A$	Discrete-gradient operator
Π_A^T		Discrete-divergence operator
Ω_A	$M \times R$ cycle matrix of $\mathcal{G}_A$	Discrete-curl operator
P	Nodes number of $\mathcal{G}_B$	Number of selected pixels
N	Edges number of $\mathcal{G}_B$	
L	Dimension of the cycle space of $\mathcal{G}_B$	
Π_B	$N \times P$ incidence matrix of $\mathcal{G}_B$	Discrete-gradient operator
Π_B^T		Discrete-divergence operator
Ω_B	$N \times L$ cycle matrix of $\mathcal{G}_B$	Discrete-curl operator

Table 1. Adopted notation

First of all, we consider the absolute phase relevant to the multichannel SAR acquisition as a node variable pertinent to both the graphs $\mathcal{G}_A$ and $\mathcal{G}_B$; by using a matrix representation, this information can be conveniently arranged in a $Q \times P$ matrix $\boldsymbol{\Phi}$ as follows:

$$\boldsymbol{\Phi} = [\boldsymbol{\varphi}^1,,\ldots,\boldsymbol{\varphi}^P] = \begin{bmatrix} \boldsymbol{\varphi}_1 \\ \vdots \\ \boldsymbol{\varphi}_Q \end{bmatrix} \tag{5}$$

where $\forall p \in \{1, 2, \ldots, P\}$ $\boldsymbol{\varphi}^p \in \mathbb{R}^Q$ encodes in a vectorial manner the pth node variable relevant to the graphs $\mathcal{G}_A$; similarly, $\forall q \in \{1, 2, \ldots, Q\}$ $\boldsymbol{\varphi}_q \in \mathbb{R}^P$ encodes the qth node variable relevant to $\mathcal{G}_B$.

Widely adopted global gradient-based PhU approaches, which have historically been developed for the single-channel case, generally consist in three processing steps [1, 29]. First, an estimation of the (wrapped) phase gradient is obtained; the estimated phase gradient is then suitably corrected (in terms of 2π multiples),and subsequently integrated to attain the unwrapped (absolute) phase.

Within the formulation of the MCh-PhU problem we concern [26], a twofold estimation of the discrete gradient field is carried out onto the considered two graphs $\mathcal{G}_A$ and $\mathcal{G}_B$, as discussed in the following. The stack of the absolute interferometric phases relevant to the M (vectorized) interferograms can be formally represented through a $P \times M$ matrix denoted by

$$\Psi = [\psi^1,\ldots,\psi^M] = \begin{bmatrix} \psi_1 \\ \vdots \\ \psi_P \end{bmatrix} \tag{6}$$

wherein the P-dimensional vector ψ^m refers to the absolute phase field pertinent to the mth interferogram. Accordingly, Ψ is formally related to the absolute (unwrapped) phase matrix Φ via the discrete gradient operator Π_A :

$$\Psi^T = \Pi_A \Phi \tag{7}$$

Note also that

$$\Psi^T = [\psi_1^T,\ldots,\psi_P^T] = [\Pi_A \varphi^1,\ldots,\Pi_A \varphi^P] \tag{8}$$

By applying the discrete gradient operator Π_B to the absolute phase of each interferometric pair, we obtain

$$\Pi_B \Psi = [\Pi_B \psi^1,\ldots,\Pi_B \psi^M] \tag{9}$$

As a result, by using Eq. (7), we get

$$\Pi_B \Psi = \Pi_B \Phi^T \Pi_A^T \tag{10}$$

Second, we consider the (wrapped) interferometric phase that is uniquely defined only in the principal value range since it is obtained as the phase of a complex function. Hence, it is convenient to formally introduce the non-injective (modulo-2π) *wrapping operator* $W : \varphi \in \mathbb{R} \rightarrow mod\ (\varphi + \pi,\ 2\pi) - \pi \in [-\pi,\ \pi)$. It should be noted that the following trivial identities hold:

$$W(W(A) \pm W(B)) = W(A \pm B) = W(W(A) \pm B) = W(A \pm W(B)) \qquad \text{a}$$
$$W(A) = A + 2\pi Z \qquad \qquad \text{b} \qquad (11)$$

where A and B represent two generic matrices and Z is a suitable integer matrix. Given the stack of the unknown absolute (unwrapped) interferometric-phases Ψ, the corresponding stack of the wrapped phases $\tilde{\Psi}$ can be conveniently expressed in terms of the *wrapped discrete gradient of (wrapped) observed phase* as follows:

$$\tilde{\Psi}^{\mathrm{T}} = [W(\Pi_{\mathrm{A}} W\,(\varphi^1)),\ldots, W(\Pi_{\mathrm{A}} W\,(\varphi^P))] = W(\Pi_{\mathrm{A}} W\,(\Phi)) = W(\Pi_{\mathrm{A}} \Phi) \qquad (12)$$

where we have exploited Eq. (11b). Note also that the pth column of $\tilde{\Psi}^{\mathrm{T}}$, i.e., $\tilde{\Psi}_p^{\mathrm{T}} = W(\Pi_{\mathrm{A}} W\,(\varphi^p))$, can be regarded as an estimate of the absolute-phase discrete gradient on the graph $\mathcal{G}_{\mathrm{A}}$. It is worth remarking that the observed (multilook) interferometric phase can however be corrupted by noise [39–41], which is taken into account by considering an additive phase noise term D. Accordingly, by using Eq. (11a), we get

$$\tilde{\Psi}^{\mathrm{T}} = W(W(\Pi_{\mathrm{A}} \Phi) + D) = W(\Pi_{\mathrm{A}} \Phi + D) \qquad (13)$$

More specifically, whenever a possible spatial filtering (e.g., conventional multilooking followed by a noise-filtering step [42]) is *independently* applied to each SAR interferometric data pair, the resulting term D in Eq. (13) implies that the phase interferograms $\tilde{\Psi}^m$, with $m \in \{1, 2, \ldots, M\}$, are no more fully time consistent (in the sense of [26, 43, 44]). To clarify this point, we observe that by using Eq. (13), it turns out that

$$W\left(\Omega_{\mathrm{A}}^T \tilde{\Psi}^{\mathrm{T}}\right) = W\left(\Omega_{\mathrm{A}}^T \left(\Pi_{\mathrm{A}} + D + 2\pi Z\right)\right) = W\left(\Omega_{\mathrm{A}}^T \Pi_{\mathrm{A}} \Phi + \Omega_{\mathrm{A}}^T D\right) = W\left(\Omega_{\mathrm{A}}^T D\right) \neq 0 \qquad (14)$$

where we have used Eq. (11b) with $A = \Pi_{\mathrm{A}} \Phi + D$ and noted that $\Omega_{\mathrm{A}}^T Z$ is an integer matrix and $\Omega_{\mathrm{A}}^T \Pi_{\mathrm{A}} = 0$ (according to Eq. (3)). Eq. (14) reads as "the wrapped discrete curl of the interferometric phase on $\mathcal{G}_{\mathrm{A}}$ (i.e., pertinent to the 'temporal' domain) is generally different from zero"; it formally expresses the (modulo-2π) *cyclic inconsistency* of the multichannel interferometric phase inherent to independently filtered SAR interferograms. Note also that Eq. (14) represents, within our framework, the generalization (to a wider class of discrete settings) of the "phase triangularity" condition in ref. [44], capturing the underlying structure of the problem within a suitable matrix formalism. With reference to the mth interferometric pair, the estimated absolute interferometric-phase gradient on the graph $\mathcal{G}_B$ is then obtained by *wrapping the discrete gradient of (wrapped) interferometric-phase field*: $\mathbf{g}^m = W(\Pi_B \tilde{\psi}^m)$. Thus, by stacking the so-obtained absolute phase gradient estimations, we get the $N \times M$ matrix $G = [\mathbf{g}^1, \mathbf{g}^2, ..., \mathbf{g}^M]$, where

$$G = W(\Pi_B \tilde{\Psi}) \tag{15}$$

Finally, by substituting Eq. (13) in Eq. (15), we obtain

$$G = W\left(\Pi_B \tilde{\Psi}\right) = W\left(\Pi_B W\left(\left[\Pi_A \Phi\right]^T + D^T\right)\right) \tag{16}$$

From Eq. (16), by using Eq. (11b), we get

$$G = W\left(\Pi_B \Phi^T \Pi_A^T + \Pi_B D^T + 2\pi \Pi_B Z\right) = W\left(\Pi_B \Phi^T \Pi_A^T + \Pi_B D^T\right) \tag{17}$$

where in the last equality we have noted that $\Pi_B Z$ is also an integer matrix. It should be emphasized that, under the assumption $D = o$, the equality between Eqs. (10) and (17) holds only up to an integer matrix multiplied by 2π.

2.3. The MCh-PhU problem as constrained optimization

In this Section, the nonlinear inversion MCh-PhU problem is reformulated as a (nonlinear) constrained optimization problem. According to the presented general formulation, we introduce in the following the MCh-PhU problem as the solution of the following matrix equation:

$$\Pi_B \Phi^T \Pi_A^T + \Pi_B D^T + 2\pi K = G \tag{18}$$

where the columns of G represent the interferometric-phase *pseudo*-gradients estimated from the observed phase, and K is an (unknown) $N \times M$ integer matrix, whose columns represent the corresponding (2π-normalized) corrections to be added to the (wrapped) interferometric-phase *pseudo*-gradients in order to recover the absolute interferometric-phase discrete gradients. It is worth noting that the term pseudo-gradient is used here to emphasize that the integration of the estimated gradient is path-dependent (non-conservative behavior); the term *residues* [1] is also typically used to connote the inconsistency of the estimated phase gradient. As a matter of fact, matrix equation (Eq. (18)) describes an *ill-posed* problem, in which the data

G generally do not constrain sufficiently the problem to get a unique solution. Additional suitable constraints and *a priori* assumption have, thus, to be introduced to solve the problem. First, for restoring the *cyclic consistency* (see Section 2.1) of the estimated pseudo-gradients pertinent to the graphs $\mathcal{G}_B$ and $\mathcal{G}_A$, two corresponding sets of (equality) constraints have to be enforced, respectively. More specifically, pre-multiplying both sides of Eq. (18) by Ω_B^T and taking into account Eq. (3), we obtain

$$\Omega_B^T \left(G - 2\pi\, K \right) = 0 \tag{19}$$

Similarly, by premultiplying both sides of the transposed version of Eq. (18) by Ω_A^T and taking also into account Eq. (3), we obtain

$$\Omega_A^T \left(G - \Pi_B D^T - 2\pi\, K \right)^T = 0 \tag{20}$$

Constraints stated by Eq. (19) imply that the columns of $G - 2\pi\, K$ must lie in the null-space of Ω_B^T. Since the matrix Π_B represents a basis to span the null-space of Ω_B^T (see Eq. (3)), we may then write $G - 2\pi\, K = \Pi_B X$, where X is a new variable. Accordingly, the corrected pseudo-gradients stack $G - 2\pi\, K$ is enforced to be a stack of discrete gradients, which can thus be unambiguously integrated. Similarly, Eq. (20) implies $[G - \Pi_B D^T - 2\pi\, K]^T = \Pi_A Y$. As a result, the two sets of constraints, stated by Eqs. (19) and (20), guarantee that the solution of the problem is effective in preserving the *cyclic consistency* (curl-free) property of the corrected gradients pertaining to the graphs $\mathcal{G}_B$ and $\mathcal{G}_A$, respectively. As a matter of fact, the solution of Eq. (18) cannot be determined by using the two sets of constraints (Eqs. (19) and (20)) only; thus, the inverse problem must be first regularized [45]. The minimum-norm methods search for a global solution that minimizes a generalized error-norm associated with an optimality criterion, so incorporating prior information about the behavior of the solution [1]. Accordingly, we resort to a regularization approach using l_1-norm minimization in weighted version, as a specific case of l_p-norm general formulation. Formally, the MCh-PhU problem may be then formulated as a constrained optimization problem for the field of integer corrections:

$$\hat{K} = \arg \min_{K \in \mathbb{Z}^{N \times M}} \| K \|_{1,C} \tag{21}$$

subject to

$$\begin{cases} \Omega_A^T K^T = \Omega_A^T \left[G - \Pi_B D^T \right]^T (2\pi)^{-1} \\ \Omega_B^T K = \Omega_B^T G (2\pi)^{-1} \end{cases} \tag{22}$$

wherein

$$\|K\|_{1,C} = \sum_{m=1}^{M} \sum_{n=1}^{N} c_{nm} |k_{nm}| \tag{23}$$

represents the weighted l_1-norm [46] of the matrix K, $C = [c_{nm}]_{N \times M}$ denotes a suitable weighting matrix, and $\mathbb{Z}$ indicates the field of integer numbers. As far as the existence of an integer solution for Eqs. (21) and (22) is concerned, it should be noted that the considerations at the end of Section 2.1 apply. Since the first matrix equation in Eq. (22) includes a generally not null (unwanted) term $\Omega_A^T D$, its fulfillment deserves further discussion. Although the evaluation of $W(\Omega_A^T D)$ can be obtained according to Eq. (14), however, a full estimation for $\Omega_A^T D$ is generally not a simple task. Further discussion is provided in Section 3. The solution of the optimization problem (Eqs. (21) and (22)) is also referred to as the minimum weighted discontinuity solution (in a weighted l_1-norm sense) [1, 23]. As a matter of fact, finding the global minimum point of the problem stated by Eqs. (21) and (22) for an arbitrary pair of graphs is, in general, a difficult task. A suboptimal strategy aimed at solving Eqs. (21) and (22) consists in adopting a two-stage approach. This is, in particular, the solution strategy implemented through the *extended minimum cost flow* (EMCF) technique [26], in which the edge structure of each considered graph is usually defined via a *Delaunay* triangulation in the Euclidean plane, to take advantage from efficient solvers for minimum cost flow (MCF) problems [47, 49, 50]. We remark that the distinctive characteristic of the EMCF approach is the extensive use of the computationally efficient MCF method. Moreover, a dual-level parallel model for EMCF has also been proposed in refs. [35] and [36]. Moreover, different approaches toward full 3D phase unwrapping have recently been proposed in refs. [63] and [64].

3. Noise-filtering of multichannel SAR interferograms

In this Section, we review the basic concepts concerning the filtering of noise that corrupts a stack of multitemporal SAR interferograms. First, the noise-filtering operation for single-channel multilook interferograms is discussed; subsequently, the general framework of the multichannel noise-filtering (MC-NF) approach, which is intimately connected with the problem of multichannel phase unwrapping, is described.

3.1. Decorrelation noise in SAR interferograms

In order to introduce the problem at hand, let us first consider one single-channel SAR interferogram obtained starting from two SAR (synthetic aperture radar) images, namely, i_1 and i_2, acquired (over the same scene on Earth) at two different times, namely, t_1 and t_2, respectively. The two SAR images can be represented via two complex-valued signals, say $i(x, r)$ and $i_2(x, r)$, with x and r denoting the two independent spatial variables (with respect to azimuth and range direction, respectively) in the radar geometry. The two complex signals can be expressed as follows [29, 51]:

$$i_1(x,r) = \gamma_1(x,r)\, e^{-j\frac{4\pi}{\lambda}r} + n_1(x,r)$$
$$i_2(x,r) = \gamma_2(x,r)\, e^{-j\frac{4\pi}{\lambda}(r+\delta r)} + n_2(x,r) \tag{24}$$

where δr is the sensor-to-target slant range difference at time t_2 with respect to time t_1, and $\gamma_1(x, r)$ and $\gamma_2(x, r)$ are the corresponding (complex-valued) reflectivity functions of the illuminated scene at time t_1 and t_2, respectively. Furthermore, the two additive (noise) contributions $n_1(x, r)$ and $n_2(x, r)$ describe random quantities that are included in Eq. (24). As a result of these noise terms and of the intrinsic random nature of the two images reflectivity functions, when the two SAR images are interfered to form a so-called interferogram, i.e., when their phase difference ψ is extracted, the interferometric phase will be noisy. An important parameter influencing the quality of the retrieved interferometric phase is the (complex) *cross-correlation* factor between the two involved SAR images, which is typically defined as

$$\chi = \frac{E[i_1 \cdot i_2^*]}{\sqrt{E\left[|i_1|^2\right] \cdot E\left[|i_2|^2\right]}} = \rho e^{j\psi} \tag{25}$$

where $\rho \in [0, 1]$, $\psi \in [-\pi, \pi)$, and the asterisk denotes the conjugate complex value. Noteworthy, the cross-correlation factor (Eq. (25)) is a complex-valued term that can be decomposed in terms of amplitude ρ (i.e., $\rho = |\chi|$) and phase ψ. For interferometric SAR images, χ can be evaluated by performing spatial averaging (known as multilooking) operations on a *statistically homogeneous* area. Indeed, the symbol $E[\,]$ in Eq. (25), which is representative of the statistical expectation operation [52], can then be replaced by the spatial averaging operation. The amplitude factor ρ, which is known to as *coherence*, accounts for the similarity between the two SAR images, whereas ψ is the multilook interferometric phase. A value for the coherence that approaches zero is representative of an uncorrelated scene, whereas coherence value that is close to unity corresponds to a noise-free interferogram.

There are several causes that are responsible for coherence decrease. As matter of fact, the cross-correlation factor in Eq. (25) depends on different noise sources, and it can be conveniently factorized as follows [51]:

$$\chi = \chi_{\text{the}} \cdot \chi_{\text{tem}} \cdot \chi_{\text{spa}} \cdot \chi_{\text{dop}} \cdot \chi_{\text{mis}} \cdot \chi_{\text{vol}} \cdot e^{j\psi} \tag{26}$$

where

- χ_{the} is the contribution of the thermal noise.

- χ_{temp} accounts for the effects due to (temporal) changes in the complex-valued reflectivity function between the two passages of the radar sensor over the illuminated area. The so-called temporal decorrelation is very difficult to be statistically modeled being associated to complex modifications of the electromagnetic response of the scene: They can be induced by human activities and/or natural causes.

- χ_{spa} is the term that takes into account the fact that from one SAR image to another the same ground resolution cell is imaged from two slightly different looking angles. The change change of the looking angle, in turn, leads to a shift between the range spectra of the two SAR images, and accordingly, it causes decorrelation since the range spectra of the two interfering SAR images are only partly overlapped. It can be shown that range spectra shift depend on the perpendicular baseline of the considered SAR data pair, and there is a limit value for the perpendicular baseline (known to as *critical baseline*) for which the two range spectra are completely non overlapped (i.e., the images are definitely uncorrelated one another) [29, 51].

- The term χ_{dop} takes into account of the so-called Doppler decorrelation effects due to the fact that SAR azimuth spectra are centered on a specific frequency (Doppler Centroid). When two SAR images with considerably different Doppler Centroid values interfere, a decorrelation noise contribution arises from the imperfect overlapping of the two related azimuth spectra. Hence, in the case that the two azimuth spectra are not overlapped at all, we have $\chi_{\text{dop}} = 0$.

- χ_{mis} accounts for possible misregistration between two SAR images.

- χ_{vol} accounts for volumetric decorrelation effects [53].

The multilook operation, leading to the multilook phase ψ in Eq. (25), reduces the level of noise corrupting interferograms, although this is paid in terms of a reduction of spatial resolution of interferograms. Multilook interferometric phase can be described by using a random quantity and, accordingly, it can be characterized via the knowledge of its probability density function. It has been shown in literature [40, 41, 54] that the *probability density function* (pdf) of an L-multilook interferometric phase (with L being the number of averaging samples in the averaging window used for the estimation of the statistical average operation involved in the calculation of Eq. (25)) can be given in terms of a Gauss *hypergeometric* function:

$$p_L(\psi) = \frac{\Gamma\left(L + \frac{1}{2}\right)(1 - \rho^2)^L \beta}{2\sqrt{\pi}\,\Gamma(L)(1 - \beta^2)^{L + \frac{1}{2}}} + \frac{(1 - \rho^2)^L}{2\pi}\,_2\Gamma_1\left[L, 1; \frac{1}{2}; \beta^2\right] \tag{27}$$

where $\psi \in [-\pi, \pi)$, $\beta = \rho cos(\psi - \psi_0)$, ρ represents the coherence, $_2F_1$ denotes the Gauss hyper-geometric function, $\Gamma(\cdot)$ is the *gamma* function, and ψ_0 represents the expected "true" value of the interferometric phase. The peak of the distribution is located at $\psi = \psi_0$.

The pdf in Eq. (27) is sketched in Figure 2a for different values of L and in Figure 2b for different values of the ρ, with $\psi_0 = 0$. By observing Figure 2a, it is clear that pdfs become narrower as the number of looks L increases (as expected). This finding is extremely important because it demonstrates that the interferometric phase may be thought to be corrupted by an additive noise random signal, namely, v, that has the same pdf as in Eq. (27) but with a zero-mean expected value, i.e., we may assume as valid the following additive model for the interfero-metric noise [54]: $\psi = \psi_0 + v$. To further investigate about the statistics of multilook interfero-grams, we can observe that the validity of Eq. (27) is only restricted to the $[-\pi, \pi)$ interval. However, this restriction does not apply when the phase signal is directly derived in the complex plane instead of the real plane. In the works of Lopez (2003) [55] and Lopez and Pottier (2007) [40], a comprehensive analytical derivation of the noise statistics in the complex plane is derived. Nonetheless, the *Cramér–Rao* bound for the standard deviation of multilook phase is given by [59]

$$\sigma_v = \frac{1}{\sqrt{2L}} \frac{\sqrt{1-\rho^2}}{\rho} \qquad (28)$$

that shows that standard deviation depends on the coherence ρ and multilook factor L. Note that the phase standard deviation approaches the limit (Eq. (28)) asymptotically as the number of looks increases.

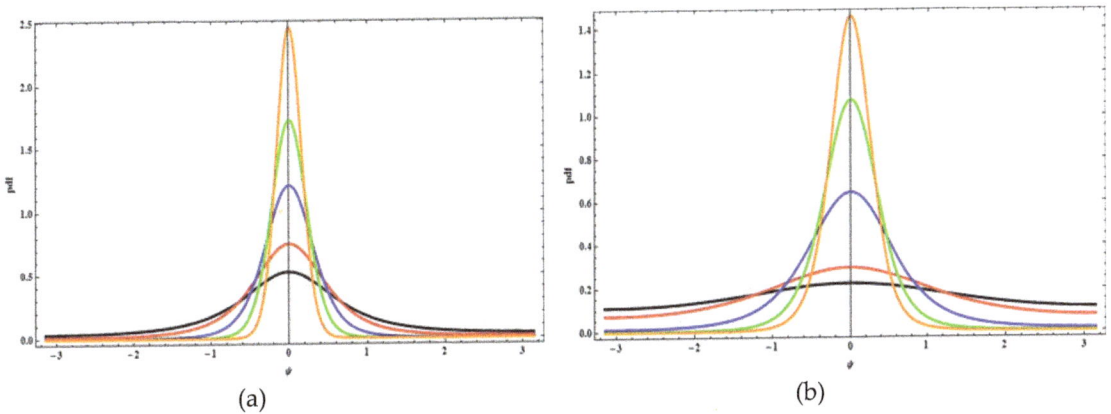

(a)   (b)

Figure 2. Probability density function of the interferometric phase ψ [rad]: (a) for different values of the number of looks L (1—black line, 2—red line, 5—blue line, 10—green line, 20—orange line), with $\rho = 0.7$; (b) for different values of the correlation coefficient ρ (0.1—black line, 0.2—red line, 0.5—blue line, 0.7—green line, 0.8—orange line) with $L = 4$.

3.2. Single-channel noise-filtering approaches for multilook interferograms

In order to mitigate the effects of decorrelation noise artifacts affecting SAR interferograms, several noise-filtering techniques, mostly working on single-channel data, have been proposed in literature over the years [42, 54, 56, 57]. As shown in previous Section 3.1, the statistics of multilook interferograms can be characterized via a probability density function expressible in closed form (Eq. (27)), and the noise standard deviation generally depends upon the *coherence* ρ and the *number of looks* L [see also Eq. (28)]. Three different multilook interferograms, which are characterized by the same perpendicular baseline (of about 100 m), have been obtained by using three SAR sensors working at the different (C, X, and L) bands of the microwave region and are depicted in Figure 3. As it is evident from Figure 3, L-band interferograms are less affected by noise than the ones pertinent to C and/or X-band.

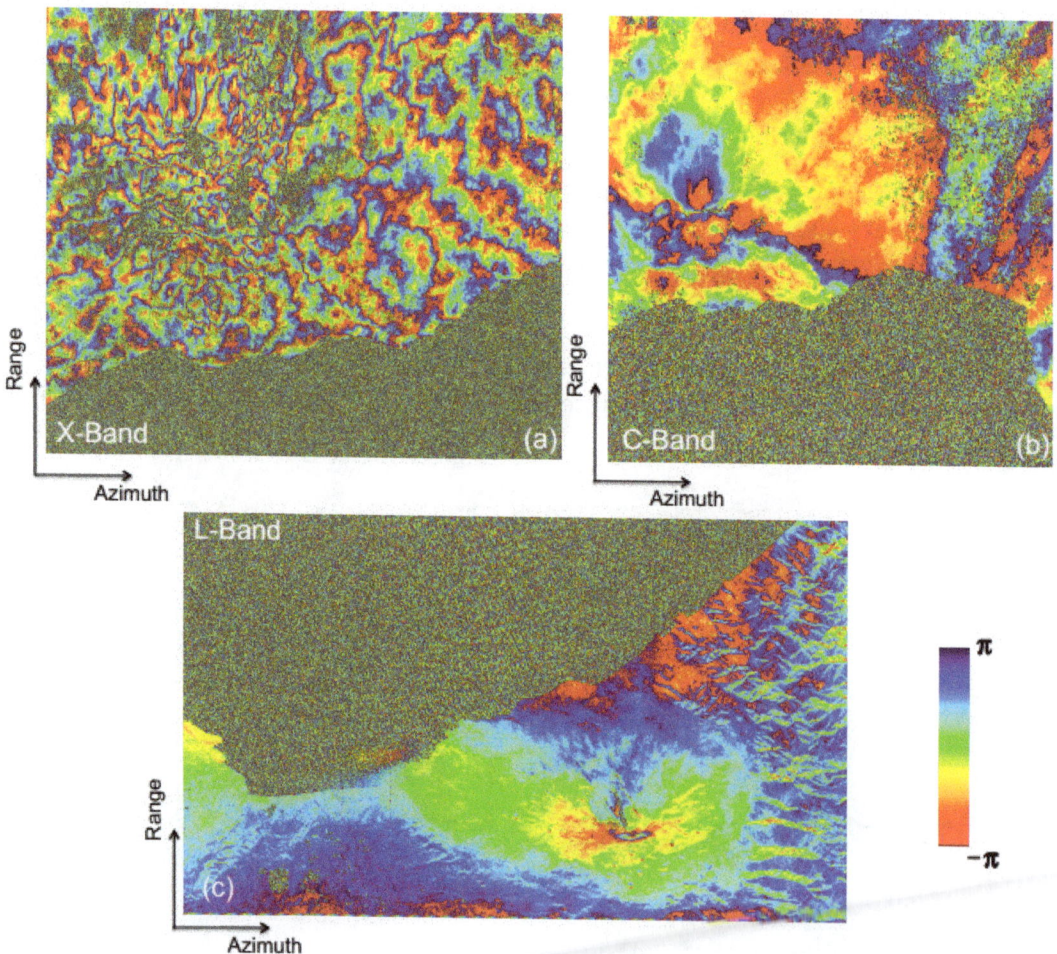

Figure 3. Multilook interferogram computed by using different SAR data pairs: (a) July 11, 2011–August 16, 2011, X-band Cosmo-SkyMed (CSK); (b) September 15, 2004–October 29, 2004, C-band ASAR/ENVISAT; (c) July 30, 2007–September 14, 2007, L-band ALOS/PALSAR-1.

It should be emphasized that coherence and noise levels can also significantly change from one SAR system to another depending on the operational wavelength.Multilook processing has been proved to be effective for noise reduction, but this is paid in terms of a decrease of the image spatial resolution. Noise filtering constitutes a preliminary step before phase unwrapping. Indeed, the multilook operation usually involves an averaging on neighboring SAR pixels, hence reducing the spatial resolution of the interferograms. Several algorithms have been proposed in literature. The most commonly used noise filter is the *boxcar* filter applied in the complex plane. Another frequently used option is provided by the *Golstein's* frequency-domain algorithm [42], which is an empirical approach proposed for geophysical applications. In this case, a complex interferogram (amplitude and phase) is segmented into overlapping rectangular patches and for each patch the relevant power spectrum Z is computed.

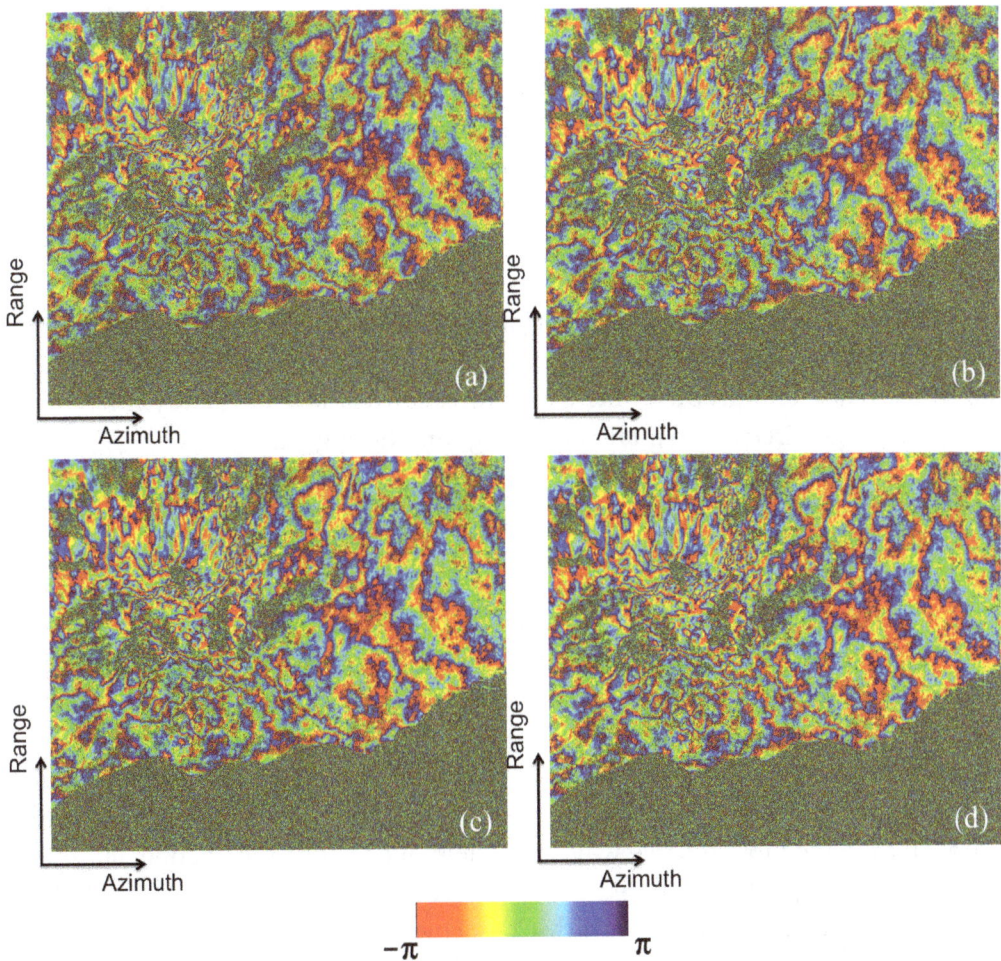

Figure 4. Multilook interferogram relevant to the SAR data pair July 11, 2011–August 16, 2011, X-band Cosmo-SkyMed (CSK): (a) Original, (b–d) Goldstein filtering, with (b) $\alpha = 0.25$, (c) $\alpha = 0.5$, and (d) $\alpha = 1.0$.

The response of the *Golstein*'s filter is then computed from the power spectrum as follows:

$$H(\xi,\eta) = |Z(\xi,\eta)|^{\alpha} \tag{29}$$

where ξ and η denote the relevant spectral variables, respectively. Notice that the filtering effect vanishes when $\alpha = 0$; conversely, the filtering effect is more pronounced as the parameter α approaches unity. We show in Figure 4 the result of the application of the Goldstein's filter to a multilook interferogram, relevant to the Mt. Etna (Italy) volcano, obtained by using the *Cosmo-Skymed* sensor of the Italian Space Agency (ASI). Specifically, different values of the filtering parameter α have been considered in Figure 4. The limited effectiveness of the filtering capabilities of Goldstein approach is evident from the result depicted in Figure 4. A modification of the Goldstein filter that relies on an adaptive choice of the filtering factor α (which depends on the spatial coherence ρ) has also (more recently) been proposed by Baran in 2003 [58]. Other filters, such as the *median* filter [59] and the two-dimensional *Gaussian* filter, are also used to reduce noise while performing multilooking operations. It is worth noting that boxcar and Goldstein filters do not adapt to the direction of the fringes because these filters are operated in a square window. In order to overcome such a limitation, Lee et al. 1998 [54] then proposed a self-adaptive filter based on local gradient slope estimation that is able to improve noise-filtering performance by exploiting directional characteristics of an InSAR interferogram. Several adaptations and relevant improvements of the *Lee filter* have subsequently proposed in literature over the recent years [56, 57], most of them based on the exploitation of the intrinsic directional behavior of InSAR interferograms. In fact, compared with the fringe phase and gradient, the fringe direction variation is gently, thus making it possible to use fringe direction to guide interferogram filtering.

3.3. The multichannel noise-filtering (MCh-NF) algorithm

The noise-filtering methods discussed in the previous Section have historically been developed to analyse and filter out the noise affecting single interferograms, separately, thus without taking into account their mutual temporal relationships. A multichannel noise-filtering problem arises when a stack of SAR interferograms need to be jointly exploited. In this case, it is profitable to develop/use noise-filtering approaches that not only exploit spatial/frequency information but can also take into account temporal relationships among available multichannel interferograms, in order to identify and filter out the noise affecting the whole interferometric data stack in the more reliable way as possible. A specific multichannel noise-filtering (MCh-NF) method [43], which is based on using a stack of time-redundant multilook interferograms, is described in this Section. The MCh-NF method is here described by adopting the same rigorous formalism and terminology used for the topological description of multichannel phase unwrapping problem presented in Section 2. According to the adopted symbolism, let us consider Q SAR images and let M be the number of multilook interferograms characterized by small perpendicular and temporal baselines.

The resulting interferometric data stack of the M (wrapped) small-baseline multilook inter-ferograms can be expressed as $\widetilde{\boldsymbol{\Psi}}^{\mathrm{T}} = [\tilde{\psi}_1^{\mathrm{T}}, \ldots, \tilde{\psi}_P^{\mathrm{T}}]$; thus, the M-dimensional vector $\tilde{\psi}_p^{\mathrm{T}}$ described the (vectorized) multichannel interferometic-phase pertinent to the pth pixel, with $p \in \{1, \ldots, P\}$ and P denoting the number of coherent pixels common to all interferograms. In particular, $\widetilde{\boldsymbol{\Psi}}$ can be expressed in terms of discrete gradient $\boldsymbol{\Pi}_A$, according to Eq. (13), as:

$$\tilde{\boldsymbol{\Psi}}^{\mathrm{T}} = W\left(\boldsymbol{\Pi}_A \boldsymbol{\Phi} + \boldsymbol{D}\right) \tag{30}$$

wherein $\boldsymbol{\Phi}$ represents the (unknown) phases associated with the available SAR images, and the matrix $\boldsymbol{D}$ describes the additive noise-term that corrupts the stack of interferograms. The noise term should be estimated and properly filter out from the generated interferograms. As discussed in Section 3.2, the term $\boldsymbol{D}$ arises since both a multilook operation and a noise-filtering procedure are typically applied to each single interferogram, separately. Both these operations are independently carried out on each single interferogram; hence, they are not necessarily time consistent. The fact that the interferograms are not fully time consistent can be formally expressed, according to Eq. (14), in terms of discrete curl $\boldsymbol{\Omega}_A^{\mathrm{T}}$, in the form:

$$W\left(\boldsymbol{\Omega}_A^{T} \tilde{\boldsymbol{\Psi}}^{T}\right) = W\left(\boldsymbol{\Omega}_A^{T} \boldsymbol{D}\right) \neq \boldsymbol{0} \tag{31}$$

which represents the *topological* generalization of the phase-triangularity condition exploited by the *SqueeSAR* technique [44]. Therefore, the *multichannel noise-filtering* (MCh-NF) approach suitably addresses the temporal inconsistencies inherent in the time-redundant multilook interferograms, which can be mathematically described in terms of the (modulo-2π) *cyclic inconsistency* of the multichannel interferometric phases [see Eq. (31)]. More specifically, MCh-NF is based on the solution of the a *nonlinear* optimization problem, as detailed in the following. First, $\forall\, p \in \{1, \ldots, P\}$, the Q-dimensional vector (Q is the number of SAR acquisitions) repre-senting the (unknown) wrapped phases $\widetilde{\boldsymbol{\Phi}}^p = W(\boldsymbol{\Phi}^p)$ is estimated as follows:

$$\hat{\tilde{\boldsymbol{\Phi}}}^p = \arg\max_{\tilde{\boldsymbol{\Phi}}^p \in \mathsf{R}^Q} \left| \overline{\boldsymbol{\zeta}}_p \circ e^{j(\tilde{\psi}_p^T - W(\boldsymbol{\Pi}_A \tilde{\boldsymbol{\Phi}}^p))} \right| \tag{32}$$

where $j = \sqrt{-1}$ denotes the imaginary unit, P denotes the number of coherent pixels to which the noise-filtering procedure is applied, the symbol $\circ$ represents the *Hadamard* product, and $\boldsymbol{\zeta_p} = [\overline{\zeta}_p^1, \cdots, \overline{\zeta}_p^M]^T$ is an M-dimensional *normalized* weighting vector representing our confidence on the quality of the exploited M (small-baseline) interferometric phases pertinent to the pth pixel, with

$$\bar{\zeta}_p^m = \frac{\zeta_p^m}{\sum_{h=1}^{M} \zeta_p^h} \tag{33}$$

wherein the generic elements ζ_p^m can be related to the *spatial coherence* as detailed after. Subsequently, these estimated vectors $\hat{\widehat{\Phi}}^p$ are used to reconstruct a new (noise-filtered) stack of time-consistent interferograms $\widetilde{\widehat{\Psi}}^T = [\widehat{\Psi}_1^T, \ldots, \widehat{\Psi}_P^T]$, where $\widehat{\Psi}_p^T = W(\Pi_A \hat{\widehat{\Phi}}^p)$ and $p \in \{1, \ldots, P\}$. We emphasize that, according to Eq. (32), the MCh-NF technique is based on searching for the (unknown) wrapped-phase vector $\widehat{\Phi}^p \in \mathbb{R}^Q$ that minimizes the (weighted) circular variance of the random (phase) vector representative of the phase difference, $\widehat{\Psi}_p^T - W(\Pi_A \widehat{\Phi}^p)$, between the "original" and the "reconstructed" interferograms.

The evaluation of the weights for the optimization problem in Eq. (32) is now addressed. Let $\breve{\Theta} = [\breve{\Theta}_{i,j}]$ be a matrix description for a generic 2-D phase map, whose corresponding vectorized representation is provided by the P-dimensional vector Θ. Each pixel of the phase map is identified by discrete range and azimuth coordinates, denoted by i and j, respectively. Accordingly, each pair (i, j) is uniquely associated with a monodimension index $p \in \{1, \ldots, P\}$ identifying an element of the vector Θ. The *spatial coherence* relevant to Θ (i.e., $\breve{\Theta}$) evaluated around the pixel (i, j) (associated with the index p) is defined as

$$\zeta_p(\Theta) = \frac{1}{(2L_R + 1)(2L_A + 1)} \left| \sum_{l=-L_R}^{L_R} \sum_{h=-L_A}^{L_A} \exp\left[j\breve{\Theta}_{i+l,j+h} \right] \right| \tag{34}$$

where $2L_R + 1$ and $2L_A + 1$ are the number of azimuth and range pixels within the used boxcar averaging window, which is centred around the generic pixel identified by the discrete range and azimuth coordinates, i and j, respectively. In particular, the mth weight ζ_p^m is expressed, according to Eq. (34), in terms of *spatial coherence* relevant to the (vectorized) interferograms $\widehat{\Psi}^m$ and evaluated around the pixel associated with the index p, in the functional form $\zeta_p^m = \zeta_p(\widehat{\Psi}^m)$, $\forall m \in \{1, \ldots, M\}$. Therefore, $\zeta_p = [\zeta_p^1, \cdots, \zeta_p^M]^T$ can be evaluated in terms of the spatial coherence directly from the stack of M multilook interferograms $\widehat{\Psi} = [\widehat{\Psi}^1, \ldots, \widehat{\Psi}^M]$. As experimentally demonstrated in ref. [43], the "reconstructed" interferograms with MCh-NF are significantly less affected by noise than the original ones. However, a group of the reconstructed interferograms, although limited, can exhibit spatial coherence values smaller than the ones relevant to the corresponding original interferograms, thus implying that a partial corruption of the spatial coherence occurs during the minimization procedure. In particular, it happens in correspondence to interferograms that were originally significantly coherent, and this is due to the fact that the strategy in Eq. (32) tends to "inject" coherence from very coherent to incoherent interferograms, by exploiting the time redundancy of the small baseline data pairs. Accordingly, in order to also preserve the spatial coherence of the very coherent interferograms, a simple nonlinear combination between the original and the reconstructed interferograms is carried out, thus further increasing the phase quality of the whole set of M reconstructed interferograms. In particular, the two sets of interferograms are combined through the following (wrapped) weighted averaging operation:

$$\bar{\psi}^m = \arctan\left(\frac{\zeta^m \circ \sin\left(\tilde{\psi}^m\right) + \hat{\zeta}^m \circ \sin\left(\hat{\tilde{\psi}}^m\right)}{\zeta^m \circ \cos\left(\tilde{\psi}^m\right) + \hat{\zeta}^m \circ \cos\left(\hat{\tilde{\psi}}^m\right)}\right) \quad \forall m \in \{1,\dots,M\} \tag{35}$$

where the symbol   represents the *Hadamard* product, and ζ^m and $\hat{\zeta}^m$ are two P-dimensional vectors. In particular, $\zeta^m = [\zeta_1^m, \cdots, \zeta_P^m]^T = [\zeta_1(\Psi^m), \cdots, \zeta_P(\Psi^m)]^T$ is expressed in terms of the *spatial coherence* relevant to the *original* multilook interferogram Ψ^m. Similarly, $\hat{\zeta}^m = [\hat{\zeta}_1^m, \cdots, \hat{\zeta}_P^m]^T = [\zeta_1(\hat{\Psi}^m), \cdots, \zeta_P(\hat{\Psi}^m)]^T$ is expressed in terms of the *spatial coherence* relevant to the *reconstructed* multilook interferogram $\hat{\Psi}^m$. The block diagram of the MCh-NF algorithm is depicted in Figure 5. A pertinent pseudo-code for computing the filtered interferometric data stack is also presented (Figure 6).

Note that the exploitation of "conventional" small baseline multilook interferograms is the distinctive characteristic of MCh-NF approach with respect to other previous solutions, such as the *SqueeSAR* [44] and *Phase Linking* [62] methods and other recently proposed multitemporal-filtering techniques [60, 61] based on constraining the analysis to distributed scatterers [29], which are identified through a pixel-by-pixel selection procedure performed at the full resolution complex SAR image spatial grid. Such a selection permits to rely on the distributed scattering hypothesis, under which the probability density function (pdf) of the complex-valued SAR image may be regarded as being a zero-mean multivariate circular normal distribution, and an appropriate maximum likelihood (ML) estimation step of the filtered phase values associated to each SAR acquisition is implemented. On the contrary, the presented MCh-NF approach focuses on conventional multilook interferograms generated without any *a priori* pixel selection step. Accordingly, in this case, it is not possible to rely on the validity of the above-mentioned distributed scattering hypothesis. Therefore, both the phase linking [62] and the phase triangulation of the SqueeSAR [44] algorithms require a preliminary identification of the statistically homogeneous pixels (SHPs) on the full-resolution range-azimuth grid. In particular, in ref. [44], the selection strategy of these pixels is based on the application of the *Kolmogorov–Smirnov* test to carefully select a homogeneous statistical population. Clearly, this requires working at the full resolution spatial scale and implies the analysis of the amplitude values of the complex SAR image pixels. A more detailed comparison among the presented MCh-NF method and the ones provided in ref. [44] can be found in ref. [43].

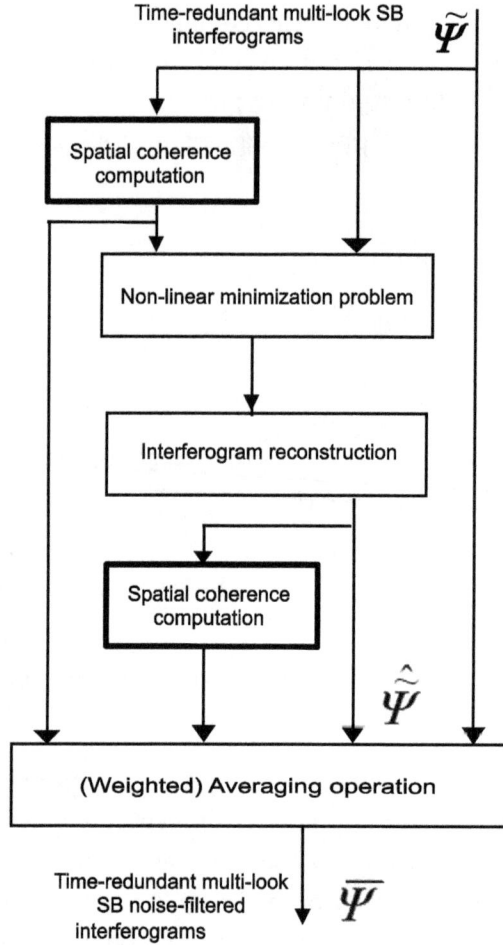

Figure 5. MCh-NF block diagram

4. Experimental results

We present in this section some results we obtained by processing a data set consisting of 39 SAR images (Track 308, Frame 2754), collected by the ENVISAT sensor between December 2002 and August 2010 over the Abruzzi region (Italy). The test-site area includes the city of L'Aquila and its surroundings, which were struck on 6 April 2009, by an Mw 6.3 earthquake that caused more than three hundred fatalities, thousands of evacuees, as well as severe industrial and residential building damages. Starting from the available SAR images, we retrieved a stack of 338 small baseline differential SAR interferograms with a maximum perpendicular baseline of 400 m and a maximum time span of 2000 days [43]. The interferograms have been computed by performing a complex multilook operation with 4 and 20 looks in the range and azimuth directions, respectively. For the interferogram generation, we used precise satellite orbit information and a three-arcsecond *shuttle radar topography mission*

(SRTM) *digital elevation model* (DEM) of the region to remove the topographic phase contributions. Finally, the multilook interferograms have been prefiltered by applying the Goldstein's filter [42].

Algorithm MCh-NF

Given: Matrix $\tilde{\Psi}^{\mathrm{T}} = [\tilde{\psi}_1^{\mathrm{T}}, \ldots, \tilde{\psi}_P^{\mathrm{T}}] = [\tilde{\psi}^1, \ldots, \tilde{\psi}^M]^{\mathrm{T}}$

{ **step 1:** Compute matrix $\hat{\tilde{\Psi}}^{\mathrm{T}} = [\hat{\tilde{\psi}}_1^{\mathrm{T}}, \ldots, \hat{\tilde{\psi}}_P^{\mathrm{T}}]$ }

for $p = 1, P$ **do**

  { **step 1.1:** Compute vectors $\bar{\zeta}_p = [\bar{\zeta}_p^1, \cdots, \bar{\zeta}_p^M]^{\mathrm{T}}$ }

  for $m = 1, M$ **do**

    Compute $\zeta_p^m = \zeta_p(\tilde{\psi}^m)$ using (34)

  end for

  for $m = 1, M$ **do**

    Compute $\bar{\zeta}_p^m$ using (33)

  end for

  Obtain vector $\hat{\bar{\Phi}}^p$ using (32)

  Compute vector $\hat{\tilde{\Psi}}_p^{\mathrm{T}} = W(\Pi_{\mathrm{A}} \hat{\bar{\Phi}}^p)$

end for

Given: Matrices $\tilde{\Psi}^{\mathrm{T}}$, $\hat{\tilde{\Psi}}^{\mathrm{T}}$ and vectors $\zeta^m = [\zeta_1^m, \cdots, \zeta_P^m]^{\mathrm{T}}$

{ **step 2:** Compute matrix $\bar{\Psi} = [\bar{\psi}^1, \ldots, \bar{\psi}^M]$ }

for $m = 1, M$ **do**

  { **step 2.1:** Compute vector $\hat{\zeta}^m = [\hat{\zeta}_1^m, \cdots, \hat{\zeta}_P^m]^{\mathrm{T}}$ }

  for $p = 1, P$ **do**

    Compute $\hat{\zeta}_p^m = \zeta_p(\hat{\tilde{\psi}}^m)$ using (34)

  end for

  Obtain vector $\bar{\psi}^m$ using (35)

end for

Figure 6. Pseudo code of the MCh-NF algorithm.

To investigate the performance of the presented noise-filtering approach, we applied the nonlinear minimization procedure in Eq. (32) to the stack of the generated (original) multilook small baseline interferograms. As a result, we retrieved a new set of noise-filtered interferograms that are characterized by generally improved coherence values. This is clearly visible in Figure 7a–f, where we compare three unfiltered (left side) interferograms with the corresponding (right side) noise-filtered interferograms. It is evident how the fringes due to the earthquake are well recovered. Such interferometric data stacks can then be used for the generation of surface deformation time series using the small baseline subset (SBAS) [6]

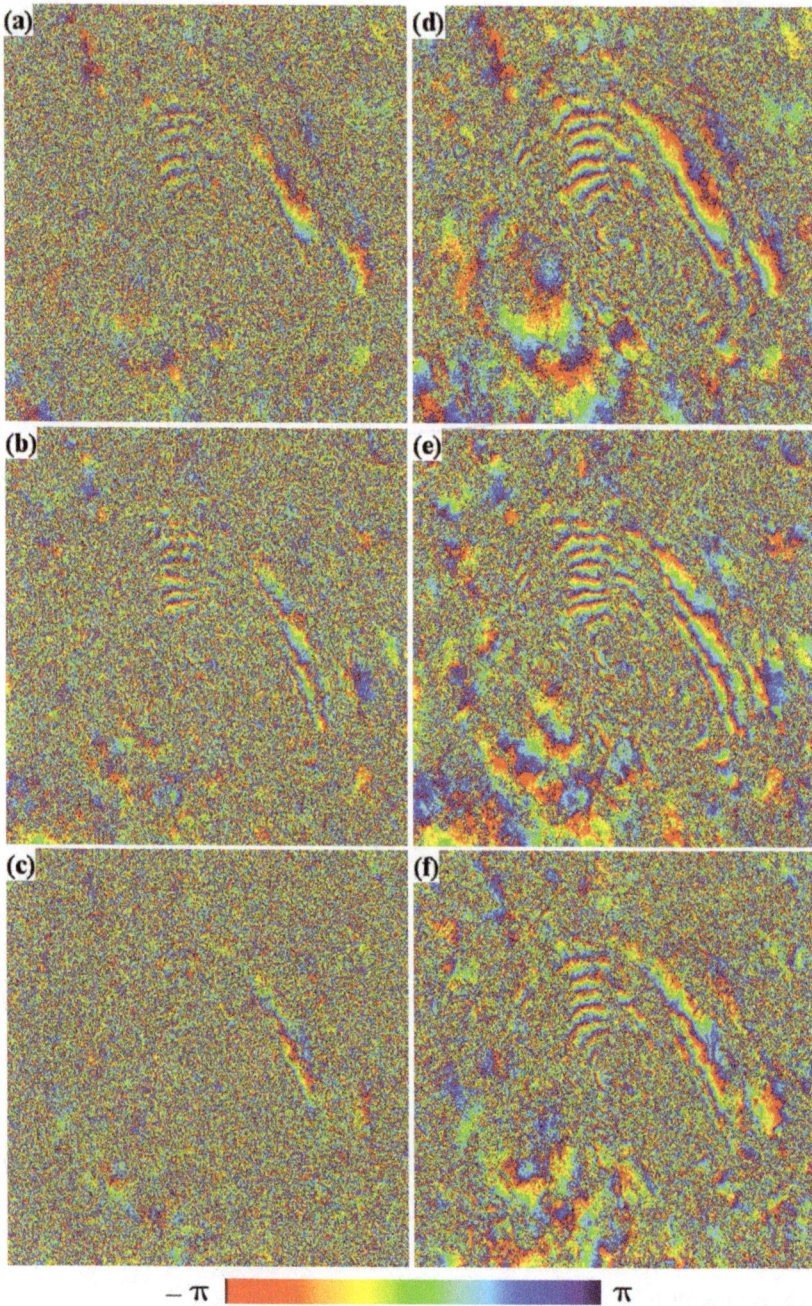

Figure 7. Comparison between the original (left column) and noise-filtered (right column) multilook interferograms relevant to the area of the Abruzzi region (Italy). (a–c) October 30, 2005, to November 8, 2009, August 21, 2005, to June 6, 2010, and August 1, 2004, to April 12, 2009, interferograms, characterized by perpendicular baseline values of 192, 145, and 395 m, respectively. (d–f) Noise-filtered multilook interferograms corresponding to the ones in panels a–c, respectively.

processing chain, whose parallel version (P-SBAS) has been proposed in refs. [13, 14, 35, 36]. This is matter for the analysis presented in the next subsection.

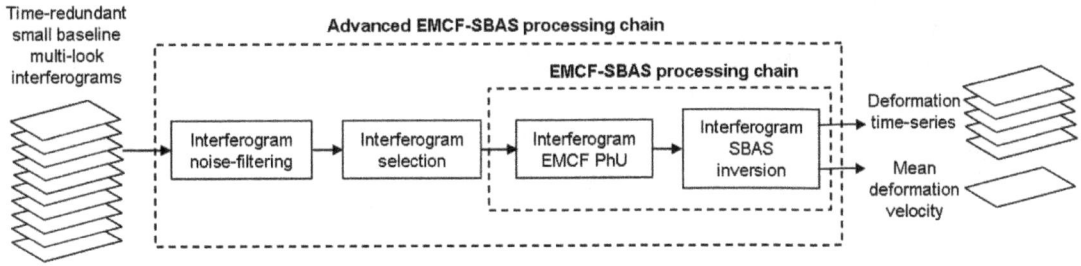

Figure 8. Block diagram of the advanced EMCF-SBAS processing chain.

4.1. The use of MCh-NF algorithm within MCh-PhU framework

We present in this subsection how the MCh-NF algorithm can be efficiently used within the SBAS processing chain, where phase unwrapping procedures are implemented through the MCh-PhU technique known as *extended minimum cost flow* (EMCF), also discussed in the first part of the chapter. Figure 8 shows the diagram block of the advanced EMCF-based SBAS processing chain [26, 35, 43], which integrates the conventional SBAS codes, exploiting the EMCF MCh-PhU procedure to perform phase unwrapping operations, with the presented MCh-NF noise-filtering technique. In addition, an effective procedure for the selection of time-redundant interferograms is also included; interested readers can find additional details in ref. [43]. To provide an example of the potential of the advanced processing chain incorporating both the discussed MCh-PhU and MC-NF techniques, we here focus on the area of *Yellow-stone* caldera, representing one of the largest and most active volcanic systems in the world. We analyze the temporal evolution of the surface deformation occurring in this area by applying the implemented EMCF-SBAS processing chain to a set of 22 ENVISAT images (Track 41, Frame 2709), acquired from May 2005 to September 2010, from which we have retrieved a corresponding set of 122 small baseline interferograms [43]. As in the previous case study (relevant to Abruzzi area), the prescribed maximum values of 400 m and 2000 days for perpendicular and temporal baseline, respectively, have also been considered. The retrieved mean deformation velocity map depicted in Figure 9, which has been obtained by applying the processing chain (MCh-NF + EMCF-SBAS) of Figure 8, allows us to recognize the complex deformation scenario affecting the Yellowstone Caldera region and its surroundings, where uplift effects and broad subsidence patterns characterize the detected deformation field. In addition, the deformation time series relevant to four pixels, whose locations are identified by the black stars labeled as P1, P2, P3, and P4, are also illustrated in Figure 9.

5. Conclusion

Within the context of SAR interferometry, two main issues related to the multichannel phase unwrapping and noise filtering for interferometric data stacks processing have been addressed. First, a rigorous gradient-based formulation for the multichannel phase unwrapping (MCh-PhU) problem has been systematically established, thus providing a topological

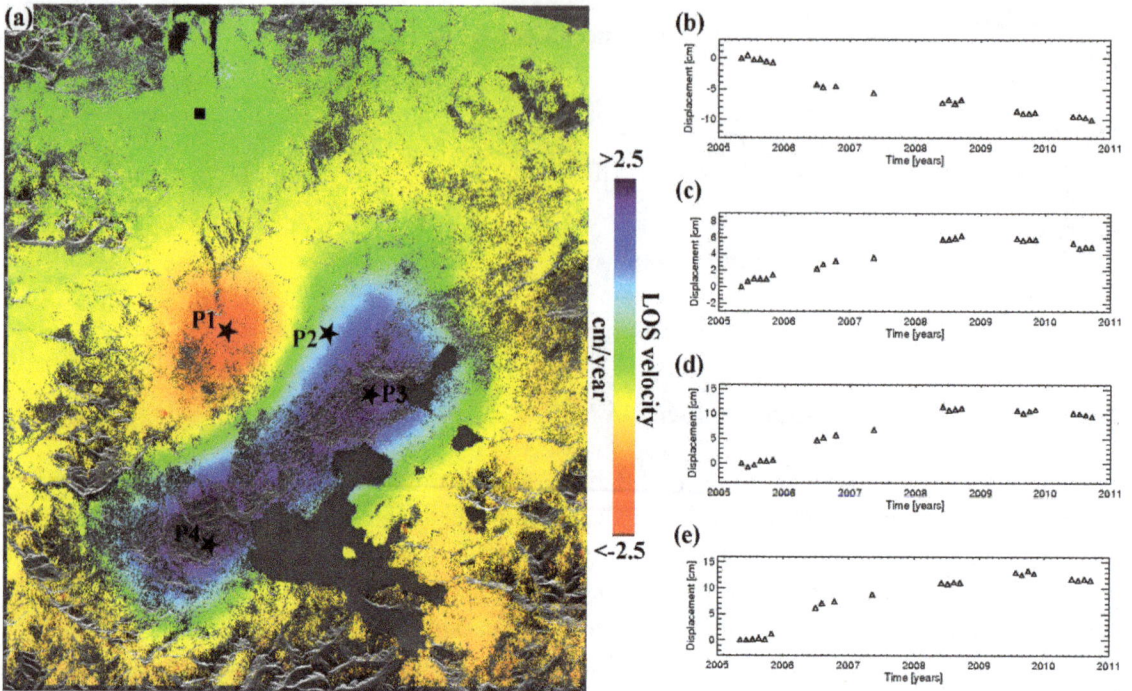

Figure 9. (a) Mean deformation velocity map (in color) of Yellowstone Caldera, computed in coherent pixels only and superimposed on the SAR amplitude image (gray-scale representation) of the zone, retrieved by applying the advanced EMCF-SBAS processing chain. The black square marks the location of the reference SAR pixel. (b–e) Deformation time series relevant to the four pixels identified via black stars in Fig. 8(a).

characterization of the problem within the purview of the theoretical foundation of the *discrete calculus*. Then the innovative MCh-NF procedure for the noise filtering of time-redundant multichannel multilook interferograms has been properly presented within the considered topological framework, by adopting a consistent formalism. Finally, some experimental results obtained with real data have been shown, thus demonstrating the effectiveness of our approaches and their relevance for geospatial phenomena analysis and understanding.

Acknowledgements

This work was supported by the Italian Ministry of University and Research (MIUR) under the project "Progetto Bandiera RITMARE." We would like to thank the European Space Agency for providing the ENVISAT ASAR data and the University of Delft, Delft, The Netherlands, for the related precise orbits. We would also like to thank Italian Space Agency (ASI), which has provided us the Cosmo-SkyMed SAR images under the framework of the European Union's Seventh Program for research, technological development, and demonstration MED-SUV project (grant no. 308665). Finally, the authors thank the Japanese Space Agency (JAXA), which has provided the used ALOS-1 data through the project entitled "Advanced Interferometric SAR Techniques for Earth Observation at L-band" (ID project 1149) in the framework of "The 4-th ALOS Research Announcement for ALOS-2" call.

Author details

Pasquale Imperatore[*] and Antonio Pepe

*Address all correspondence to: imperatore.p@irea.cnr.it

Istituto per il Rilevamento Elettromagnetico dell'Ambiente (IREA), National Research Council (CNR) of Italy, Napoli, Italy

References

[1] D. C. Ghiglia, M. D. Pritt, Two-dimensional phase unwrapping: theory, algorithms and software, New York, John Wiley, 1998.

[2] Goldstein, and H.A. Zebker, "Mappings mall elevation changes over large areas: Differentia radar interferometry," J. Geophys. Res., vol.94, no.B7, pp. 9183–9191, 1989.

[3] D. Massonnet and K. L. Feigl, "Radar interferometry and its application to changes in the Earth's surface," Rev. Geophys., vol. 36, pp. 441–500, 1998.

[4] Bürgmann, P. A. Rosen, and E. J. Fielding, "Synthetic aperture radar interferometry to measure Earth's surface topography and its deformation," Annu. Rev. Earth Planet. Sci., vol. 28, pp. 169–209, May 2000.

[5] A. Ferretti, C. Prati, and F. Rocca, "Permanent scatterers in SAR interferometry," IEEE Trans. Geosci. Remote Sens., vol. 39, no. 1, pp. 8–20, Jan. 2001.

[6] P. Berardino, G. Fornaro, R. Lanari, and E. Sansosti, "A new algorithm for surface deformation monitoring based on small baseline differential SAR interferograms," IEEE Trans. Geosci. Remote Sens., vol.40, no.11, pp. 2375–2383, Nov.2002.

[7] A. Hooper, H. Zebker, P. Segall, and B. M. Kampes, "A new method for measuring deformation on volcanoes and other natural terrains using InSAR persistent scatterers," Geophys. Res. Lett., vol. 31, no. 23, p. L23 611, Dec. 2004, DOI: 10.1029/2004GL021737.

[8] M. Crosetto, B. Crippa, and E. Biescas, "Early detection and in-depth analysis of deformation phenomena by radar interferometry," Eng. Geol., vol. 79, no. 1/2, pp. 81–91, Jun. 2005.

[9] B. M. Kampes, "Radar Interferometry: Persistent Scatterer Technique," Springer, 2006.

[10] A. Hooper and H. Zebker, "Phase unwrapping in three dimensions with applications to InSAR time series," J. Opt. Soc. Am. A, vol. 24, no. 9, pp. 2737–3747, Aug. 2007.

[11] J. Hunstad, A. Pepe, S. Atzori, C. Tolomei, S. Salvi, and R. Lanari, "Surface deformation in the Abruzzi region, Central Italy, from multi-temporal DInSAR analysis," Geophys. J. Int., vol. 178, no. 3, pp. 1193–1197, Sep. 2009.

[12] S. Elefante, P. Imperatore, I. Zinno, M. Manunta, E. Mathot, F. Brito, J. Farres, W. Lengert, R. Lanari, F. Casu, "SBAS-DINSAR time series generation on cloud computing platforms," Proc. IEEE IGARSS 2013, pp. 274–277, Melbourne (AU), July 2013.

[13] P. Imperatore, et al., "Scalable performance analysis of the parallel SBAS-DINSAR algorithm," Proc. IEEE IGARSS 2014, pp. 350–353, Québec City, Canada, July 2014.

[14] F. Casu, S. Elefante, P. Imperatore, I. Zinno, M. Manunta, C. De Luca, R. Lanari, "SBAS-DInSAR parallel processing for deformation time series computation," IEEE J. Select. Topics Applied Earth Observ. Remote Sens., vol.7, no.8, pp. 3285–3296, Aug. 2014.

[15] R. Gens, "Two-dimensional phase unwrapping for radar interferometry: Developments and new challenges," Int. J. Remote Sens., vol.24, N.4, pp. 703–710, 2003.

[16] C. W. Chen and H. A. Zebker, "Phase unwrapping for large SAR interferograms: statistical segmentation and generalized network models," IEEE Trans. Geosci. Remote Sens., vol. 40, No. 8, pp-1709-1719, Aug. 2002.

[17] M. Costantini, "A novel phase unwrapping method based on network programming," IEEE Trans. Geosci. Remote Sens., vol. 36, pp. 813–821, May 1998.

[18] C. W. Chen and H. A. Zebker, "Network approaches to two-dimensional phase unwrapping: intractability and two new algorithms," J. Opt. Soc. Am., vol.17, no. 3, pp. 401–414, Mar. 2000.

[19] K. Zhang, L. Ge, Z. Hu, A. Hay-Man Ng, X. Li, and C. Rizos, "Phase unwrapping for very large interferometric data sets," IEEE Trans. Geosci. Remote Sens., vol. 49, No. 10, pp. 4048–4061, Oct. 2011.

[20] W. Xu and I. Cumming, "A region-growing algorithm for InSAR phase unwrapping," IEEE Trans. Geosci. Remote Sens., 37, pp. 124–134. 1999.

[21] G. F. Carballo and P. W. Fieguth, "Hierarchical network flow phase unwrapping," IEEE Trans. Geosci. Remote Sens., vol. 40, No. 8, pp. 1695–1708, Aug. 2002.

[22] G. Fornaro, A. Pauciullo,D. Reale, "A null-space method for the phase unwrapping of multitemporal SAR interferometric stacks," IEEE Trans. Geosci. Remote Sens., vol. 49, no.6, pp. 2323–2334, June 2011.

[23] T. J. Flynn, "Two-dimensional phase unwrapping with minimum weighted discontinuity," J. Opt. Soc. Am. A, 14(10), pp. 2692-2701. 1997.

[24] O. Mora, J. J. Mallorquí, and A. Broquetas, "Linear and nonlinear terrain deformation maps from a reduced set of interferometric SAR images," IEEE Trans. Geosci. Remote Sens., vol. 41, no. 10, pp. 2243–2253, Oct. 2003.

[25] S. Usai, "A least squares database approach for SAR interferometric data," IEEE Trans. Geosci. Remote Sens., vol. 41, no 4, pp. 753–760, April 2003.

[26] A. Pepe, and R. Lanari, "On the extension of the minimum cost flow algorithm for phase unwrapping of multitemporal differential SAR interferograms," IEEE Trans. Geosci. Remote Sens., vol. 44, no. 9, pp. 2374–2383, Sept. 2006.

[27] A. P. Shanker and H. Zebker, "Edgelist phase unwrapping algorithm for time series InSAR analysis," J. Opt. Soc. Am. A, vol. 27, no. 3, pp. 605–612, Mar. 2010.

[28] M. Costantini, S. Falco, F. Malvarosa, F. Minati, F. Trillo, and F. Vecchioli, "A general formulation for robust integration of finite differences and phase unwrapping on sparse multidimensional domains," in Proc. Fringe, Frascati, Italy, Dec. 2009.

[29] R. Bamler, P.Hartl, "Synthetic aperture radar interferometry," Inverse Problems, vol. 14, no.4, R1, 1998.

[30] L. J. Grady, J. Polimeni, Discrete Calculus: Applied Analysis on Graphs for Computational Science, Springer, 2010.

[31] N. Biggs, Algebraic Graph Theory. Cambridge University Press, Cambridge, UK, 1994.

[32] C. Berge, Graphs and Hypergraphs. North-Holland Publishing Co., Amsterdam, 1973.

[33] R. Diestel, Graph Theory, Springer-Verlag, New-York, 2000.

[34] L. Grady, "Minimal surfaces extend shortest path segmentation methods to 3D," IEEE Trans Pattern Anal. Mach. Intell., vol. 2, no. 32, pp. 321–334, 2010.

[35] P. Imperatore, A. Pepe, R. Lanari, "Multichannel phase unwrapping: problem topology and dual-level parallel computational model," IEEE Trans. Geosci. Remote Sens., vol. 53, no.10, pp. 5774–5793, October 2015.

[36] P. Imperatore, A. Pepe, R. Lanari, "High-performance parallel computation of the multichannel phase unwrapping problem," Proceedings of the IEEE International Geoscience and Remote Sensing Symposium, IGARSS 2015, Milan, Italy, July 2015.

[37] P. A. Rosen, S. Hensley, I. R. Joughin, F. K. Li, S. R. Madsen, E. Rodriguez, and R. M. Goldstein, "Aperture radar interferometry," Proc. IEEE, vol. 88, 3,pp. 333–381, 2000.

[38] H. A. Zebker and J. Villasenor, "Decorrelation in interferometric radar echoes," IEEE Trans. Geosci. Remote Sens., vol. 30, pp. 950–959, Sept. 1992.

[39] C. H. Gierull, "Statistical analysis of multilook SAR interferograms for CFAR detection of ground moving targets," IEEE Trans. Geosci. Remote Sens., vol. 42, n. 4, April 2004.

[40] C. Lopez-Martinez and E. Pottier, "On the extension of multidimensional speckle noise model from single-look to multilook SAR image, "IEEE Trans. Geosci. Remote Sens., vol. 45, n. 2, February 2007.

[41] Lee, J. S. K. W. Hopple, S. A. Mango and R. Miller: "Intensity and phase statistics of multilook polarimetric interferometric SAR imagery," IEEE Trans. Geosci. Remote Sens., 32(5), 1017–1028, 1994.

[42] R. M. Goldstein, and C. L. Werner, "Radar interferogram filtering for geophysical applications," Geophys. Res. Lett., vol. 25, pp. 4035–4038, 1998.

[43] A. Pepe, Y. Yang, M. Manzo, R. Lanari, "Improved EMCF-SBAS processing chain based on advanced techniques for the noise-filtering and selection of small baseline multi-look DInSAR interferograms," IEEE Trans. Geosci. Remote Sens., vol. 53, no. 8, pp. 4394–4417, Aug. 2015.

[44] A. Ferretti, A. Fumagalli, F. Novali, C. Prati, F. Rocca, and A. Rucci, "A new algorithm for processing interferometric data-stacks: SqueeSAR," IEEE Trans. Geosci. Remote Sens., vol. 49, pp. 3460–3470, Sept. 2011.

[45] S. S. Rao, Engineering Optimization: Theory and Practice, Fourth Edition, John Wiley & Sons, Inc, 2009.

[46] K. Yosida, Functional Analysis, Berlin, Germany: Springer-Verlag, 1980.

[47] D. Bertsekas, P. Tseng, "The relax codes for linear minimum cost network flow problems," Ann. Oper. Res., V. 13, 1988.

[48] M. Costantini, P.A. Rosen, "A generalized phase unwrapping approach for sparse data," Proc. IGARSS99, pp. 267–269, Hamburg (Germany), 1999.

[49] R.K. Ahuja, T.J. Magnanti, J.B. Orlin, Network Flows: Theory, Algorithms, and Applications, Prentice Hall, Ney Jersey, 1993.

[50] D. Bertsekas and P. Tseng. RELAX-IV: a faster version of the RELAX code for solving minimum cost flow problems. Technical report. Department of Electrical Engineering and Computer Science, MIT, Cambridge, MA, 1994.

[51] G. Franceschetti and R. Lanari, Synthetic Aperture Radar Processing Boca Raton, FL: CRC, Mar. 1999.

[52] H. Stark and J. W. Woods, Probability and Random Processes with Applications to Signal Processing, 3rd edition, Pearson, 2012.

[53] C. Elachi, "Spaceborne radar remote sensing: applications and techniques. Institute of Electrical and Electronics Engineers, 1998.

[54] Jong-Sen Lee, Konstantinos P. Papathanassiou, Thomas L. Ainsworth, Mitchell R. Grunes, and Andreas Reigber, "A new technique for noise filtering of SAR interfero-

metric phase images," IEEE Trans. Geosci. Remote Sens., vol. 36, no. 5, pp. 1456–1465, Sep. 1998.

[55] C. López-Martínez and X. Fàbregas, "Polarimetric SAR speckle noise model," IEEE Trans. Geosci. Remote Sens., vol. 41, no. 10, pp. 2232–2242, Oct. 2003.

[56] Sihua Fu, Xuejun Long, Xia Yang, and Qifeng Yu, "Directionally adaptive filter for synthetic aperture radar interferometric phase images," IEEE Trans. Geosci. Remote Sens., vol. 51, no. 1,Jan 2013.

[57] Qingsong Wang, Haifeng Huang, Anxi Yu, and Zhen Dong, "An efficient and adaptive approach for noise filtering of SAR interferometric phase images," IEEE Trans. Geosci. Remote Sens. Lett., vol. 8, no. 6, Nov 2011.

[58] I. Baran, M. P. Stewart, B. M. Kampes, Z. Perski, and P. Lilly, "A modification to the Goldstein radar interferogram filter," IEEE Trans. Geosci. Remote Sens., vol. 41, no. 9, Sept. 2003.

[59] E. Rodriguez, J. M. Martin, "Theory and design of interferometric synthetic aperture radars," IEE Proceedings F Radar and Signal Processing, vol.139, no.2, pp. 147–159, Apr 1992.

[60] A. Parizzi, and R. Brcic, "Adaptive InSAR stack multilooking exploiting amplitude statistics: a comparison between different techniques and practical results," IEEE Trans. Geosci. Remote Sens. Lett., 8, pp. 441–445, May 2011.

[61] G. Fornaro, D. Reale and S. Verde, "Adaptive spatial multilooking and temporal multilinking in SBAS interferometry," Proceedings of the IEEE International Geoscience and Remote Sensing Symposium (IGARSS), Munich (Germany), July 2012.

[62] A. Parizzi and R. Brcic, "Adaptive InSAR stack multi-looking exploiting amplitude statistics: a comparison between different techniques and practical results," IEEE Trans. Geosci. Remote Sens. Lett., vol. 8, no. 3, pp. 441–445, May 2011.

[63] A. P. Shanker and H. Zebker, "Edgelist phase unwrapping algorithm for time series InSAR analysis," J. Opt. Soc. Am. A, Opt. Image Sci. Vis., vol. 27, no. 3, pp. 605–612, Mar. 2010.

[64] M. Costantini, S. Falco, F. Malvarosa, F. Minati, F. Trillo, and F. Vecchioli, "A general formulation for robust integration of finite differences and phase unwrapping on sparse multidimensional domains," in Proc. Fringe, Frascati, Italy, Dec. 2009.

5

Remote Sensing-Based Biomass Estimation

José Mauricio Galeana Pizaña, Juan Manuel Núñez Hernández and
Nirani Corona Romero

Additional information is available at the end of the chapter

Abstract

Over the past two decades, one of the research topics in which many works have been
done is spatial modeling of biomass through synergies between remote sensing, forestry, and ecology. In order to identify satellite-derived indices that have correlation
with forest structural parameters that are related with carbon storage inventories and
forest monitoring, topics that are useful as environmental tools of public policies to
focus areas with high environmental value. In this chapter, we present a review of different models of spatial distribution of biomass and resources based on remote sensing that are widely used. We present a case study that explores the capability of
canopy fraction cover and digital canopy height model (DCHM) for modeling the
spatial distribution of the aboveground biomass of two forests, dominated by *Abies
Religiosa* and *Pinus* spp., located in Central Mexico. It also presents a comparison of
different spatial models and products, in order to know the methods that achieved the
highest accuracy through root-mean-square error. Lastly, this chapter provides concluding remarks on the case study and its perspectives in remote sensing-based biomass estimation.

Keywords: Aboveground biomass, forest, remote sensing, modeling, resources

1. Introduction

Forest ecosystems are about 31% of the total land cover of the earth [1], being one of the most
important ecosystems due to economic goods and environmental services they provide. One
of these services is as an environmental regulator, reducing the concentration of carbon dioxide
(greenhouse gas) from the atmosphere and transforming it into oxygen and biomass through
photosynthesis, thereby playing an important role in the global carbon cycle [2–4].

Biomass is defined as the dry weight of both aboveground biomass (AGB) and belowground biomass (BGB) living mass of vegetation, such as wood, bark, branches, twigs, stumps, or roots as well as dead mass of litter associated with the soil [4–6]. According to this, it can be considered as a measure of forest structure and function. Thus, by knowing the spatial distribution of biomass, it is possible to calculate the net flow of terrestrial carbon, nutrient cycling, forest productivity, biomass energy, and carbon storage and sequestration by the forest, reducing the uncertainty of carbon emission and sequestration measures to support climate change modeling studies [5–8].

Since calculating field measures of BGB is difficult, most studies have focused on calculating AGB. The most accurate way to obtain AGB data is by using field measurements and allometric equations for individual trees; however, these techniques are difficult to implement because they are time consuming and labor intensive. Furthermore, forests are a complex and widely distributed ecosystem, which makes these techniques expensive to apply in large areas; therefore, they cannot provide the spatial distribution of biomass [4,6].

An alternative form to map and monitor spatial distribution of AGB is through the use of remote sensing-based techniques, because through them it is possible to obtain a continuous and repetitive collection of digital data from the same area with different spatial resolutions, covering large areas and reducing processing time and costs [5,6]. Due to the increasing availability of satellite imagery, several researches have been developed to prove the effectiveness of both imagery data provided by different sensors and diverse modeling approaches [4,6,9]. In order to estimate AGB, two main kinds of models have been used: direct and indirect. The direct models measure biomass throughout the relationship between spectral data response and biomass field measurements. For the indirect models, biomass is estimated from biophysical parameters or forest structural metrics [10].

This chapter reviews the main models used for estimating biomass and key resources used in remote sensing (Sections 1 and 2). The case study integrates some models and resources applied in forests dominated by *Abies Religiosa* and *Pinus* spp. located in Central Mexico. In the last section, concluding remarks are provided about the best biomass estimation models as well as their limitations.

2. Biomass modeling

Several factors can affect the remote sensing-based AGB estimation, such as insufficient sample data, atmospheric conditions, complex biophysical environments, scale of the study area, availability of software, spatial resolution of remotely sensed data, or mixed pixels, among others [6, 10]. In order to introduce different approaches that have been developed to reduce the uncertainties produced by these factors on the estimation of the spatial distribution of AGB, the most commonly used models will be described in this section.

2.1. Field measurements and allometric equations

All spatial distribution AGB estimation models need high quality and representative field data to be implemented. Therefore, forest inventories are the most common approach to obtain detailed and periodic data. They are held for monitoring, modeling, and predicting several biophysical processes, such as stocking levels, harvests, diseases, and pests, among others. Therefore, they are generally implemented at several scales to obtain different structural parameters, either through the data aggregated from stand-level management inventories or by plots established through a statistical sampling design [3,4,6,11].

Since the grouping of data from stand-level management inventories tend to underestimate the forested areas and stock volume calculation, presently in most of the countries a statistical sampling design is used. In this procedure, the sampling plots are randomly selected from a population where each one of them has the same positive known probability to be chosen. They can be selected by different methods. One of them is a systematic sampling procedure based on the use of grids of randomly selected points in two dimensions, with a 0.5–20-km separation range. In other cases, plot locations are randomly selected by regular polygons created by a tessellation of large areas, which can be stratified when different sampling intensities are required, for example, when different land uses and covers exist [11].

In order to make a more efficient sampling, a plot generally conforms clusters, commonly of four plots but, in some cases, they can be as large as 18, with a huge variety of shapes and sizes. Circular plots are commonly used in Boreal and Temperate forests, whereas square and rectangular ones are used in rainforests. Plots can be combined with other sampling methods like transect for measuring deadwood or soil pits to assess soil carbon [11].

Commonly, the information gathered about a plot is location, number of trees, species, health, and site description, among others. In addition, individual tree dendrometric variables are considered, such as diameter at breast height (DBH), tree height, crown size, and canopy cover, DBH and tree height being the most commonly used parameters to derivate AGB through allometric equations [3,4,10].

The allometric biomass equation is the most common and accurate method to translate forest inventory reports of individual tree data to tree and stand biomass. It is a mathematical relation between total tree biomass (stem, branch, and foliage) and its DBH or both DBH and height, applying a least-squares regression of logarithmic equation. It can be both species- or site specific (e.g., *Pinus montezume* or *Abies religiosa*) or more generic (e.g., pine gender or tropical hardwoods); however, it has been observed that biomass equations at the site-specific level produce better results than generic equations [3,8,12].

The most accurate method to obtain allometric equations is a destructive process in which individual trees from a wide range of DBH, distributed in a local forest, must be felled and separated into boles, branches, and leaves. Then boles are cut into sections and weighted in the field as leaves and branches. After that, a thick disc sample must be cut from the base of each bole section and a subsample is extracted for the other side of the disk, both being dried at 105°C until a constant weight is reached to obtain the dry mass, to estimate the moisture content and the wood density. The total biomass therefore is the sum of the dry mass of the branches, the leaves, and the various sections of the stem [3,13].

Since it has been observed that growing plants maintain the weight proportion between different parts, it is possible to build a nonlinear mathematical model to relate biomass with DBH using these parameters [12]. One of the most common models used is the logarithmic equation (1):

$$Ln(B) = Ln(a) + bLn(D) \qquad (1)$$

where B is the biomass, a and b are scalar coefficients estimated by least-squares linear regression, and D is the DBH.

Using allometric equations, it is possible to compute the total AGB for a given area using biomass expansion factors or conversion tables; however, these approaches do not provide the spatial distribution of AGB [3,10]. Later in this chapter, different models created to obtain spatial distribution of AGB are explained.

2.2. Regression models

One of the most common methods to estimate biomass is the regression analysis, which is a statistical technique to investigate and model the relationship between variables. Traditionally, in the remote sensing approach, the regression analysis techniques applied to AGB estimation are based on the quantitative relationship between ground-based data and satellite information, such as spectral reflectance, radar, or light detection and ranging (LiDAR) data [6,14,15]. Models based on regression analysis are considered to be relatively easy to implement and can provide accurate results through their application at all spatial scales [16]. Generally, these methodologies consist of three major steps: biomass estimation based on fieldwork, establishment of regression model between field biomass and satellite information of corresponding pixels, and the use of regression models to generate a biomass image with the spatial prediction.

These remote sensing-based biomass estimation methods assume that the forest information, obtained by the sensors, is highly correlated with AGB. According to this, the keys for biomass estimation are the use of appropriate variables and the development of suitable estimation models for sufficient sample plots, using regression methods that aim for an efficient integration of multisource data, necessary to get better biomass estimation. Spectral data, radar, and LiDAR have their own positive and negative characteristics and proper integration of them can improve biomass estimation accuracy [17].

After data integration, the correct use of regression methods for establishing biomass estimation models is also important. Many models have been developed based on multiple combinations of in situ tree parameters calculated through linear regression (LR) or nonlinear regression (NLR) models [18–20]. Multiple regression analysis may be the most frequently used approach for developing biomass estimation models [21–23]. In both cases, these parametric algorithms assume that the relationships between dependent (e.g., biomass) and independent (derived from remote sensing data) variables have explicit model structures that can be specified a priori by parameters [15]. Generally, the independent variables can be spectral bands, vegetation indices, textural images, LiDAR height, and synthetic aperture radar (SAR) backscatter; in some cases, the use of subpixel information offered better estimation results than per-pixel-based spectral signatures [24].

However, the linear regression approach has been known to mislead the prediction of the studied variable at values beyond a saturation point of the canopy reflectance [25]. Since biomass is usually nonlinearly related to remote sensing variables, nonlinear models such as power models [26], logistic regression models [27], and geographically weighted regression models [16] were often used to estimate biomass with more accuracy. Nonetheless, some estimation methods have been established as a nonparametric alternative to the use of regression approaches for biomass modeling: k-nearest neighbor (k-NN), artificial neural network (ANN), regression tree, random forest, support vector machine (SVM), and maximum entropy (MaxEnt) [15].

2.3. Geostatistical models

Some studies have used geostatistics as the main approach for biomass forest estimation in order to predict variables related to forest structure (e.g., tree height and volume) and aboveground biomass and carbon measurements in unsampled sites based on known values of adjacent spatial data as forest inventory sites [28,29]. Other recent works have explored the synergy between geostatistical models with remotely sensed data to improve estimations using remote sensing indices as spatial secondary variables [10, 16, 30–32].

Geostatistics was defined in the 1960s by Georges Matheron, who generalized a set of techniques developed by Krige (1951) in order to exploit the spatial correlation to make predictions in evaluating reserves of gold mines in South Africa. This generalization is detailed in his regionalized variable theory in 1970 [33]. The purpose of geostatistics is the estimation, prediction, and simulation of the values of a variable that is distributed through space [34]. This theory assumes that a variable measured in a spatial domain corresponds to a random variable $z(x)$, assuming that the structure of the phenomenon having spatial correlation is considered a regionalized variable; therefore, a set of spatially distributed random variables will be a random function $Z(x)$. This provides the theoretical basis for establishing the spatial structural characteristics of natural phenomena. Moreover, it can be used as a tool for calculating the value of a variable in a certain position in space, knowing the values of that variable among adjacent positions in space, which is known as interpolation [35].

There are diverse methods of interpolations, which can be classified into two main groups: deterministic and geostatistical. The deterministic techniques are based directly on some properties of similarity of adjacent measured values (e.g., distance), which establish a set of mathematical formulas that determine the smoothness of the resultant surface interpolated. Examples of these are the inverse distance weighting (IDW), nearest neighbors, splines, and triangular irregular network (TIN) [35]. Geostatistical techniques studied spatial autocorrelation of the variables in order to fit a spatial dependence model to a set of random variables.

This approach produces predictions and also generates an error surface concerning the uncertainty-associated analyzed model [33].

The spatial dependence is the spatial behavior of a phenomenon, derived from spatial patterns in terms of distance and similarity and/or contrast of a spatial unit or relative spatial location to other spatial units [36]. It has its basis in Tobler's first law of geography proposed in 1970 that says, "everything is related to everything else, but near things are more related than distant things" [37].

The kriging algorithms are one such example, which is mostly used in geosciences, ecology, and geomatics [35]. Kriging is a generic name for a family of generalized least-squares regression techniques, where the spatial structural characteristics are accomplished by the semivariogram function as a metric of the spatial autocorrelation [33].

All kriging estimators are variants of the following basic equation (2):

$$\hat{Z}(x_0) - \mu = \sum_{i=1}^{n} \lambda_i \left[Z(x_i) - \mu(x_0) \right] \tag{2}$$

where μ is a known stationary mean, assumed to be constant over the whole domain and calculated as the average of the data. The parameter λ is the kriging weight; n is the number of sampled data points used to make the estimation, and $\mu(x_0)$ is the mean of the samples within the search window [33].

Different studies have applied univariate and bivariate geostatistical interpolations in order to calculate the forest volume [28], aboveground biomass [10,16,29–31], and carbon in the aboveground biomass [32]. The most commonly used technique for univariate-based modeling is kriging [28], whereas in bivariate-based modeling regression-kriging [10,30–32], cokriging [31], kriging with external drift [29], cokriging regression [31], and geographically weight regression (GWR) are used [16].

2.4. Nonparametric models

Similar to regression models, nonparametric algorithms are based on the use of different sensor data, for example, spectral, radar, and LiDAR [21,38], using many of these models in the forest attributes estimation [39–41]. They are a framework for creating complex nonlinear biomass models based on the use of remote sensing variables and as alternatives for the parametric approaches. Common nonparametric algorithms include k-nearest neighbor (k-NN), artificial neural network (ANN), random forest, support vector machine (SVM), and maximum entropy (Max Ent).

One of the most applied nonparametric methods is the nearest neighbor approach (NN). In the context of forest attribute estimation, the NN methods have been first introduced in the late 1980s [42]. In the NN methods, the value of the target variable at a certain location is predicted as a weighted average of the values of neighboring observations, with k-nearest neighbors spectrally using a weighting method [43]. Several methods have been offered to measure the distance from the target unit to the neighbors. In the NN approach, the choice of the k value, type of distance measure including Euclidean and Mahalanobis, along with weighed function are the critical factors influencing the estimation accuracy [44,45].

The ANN provides a more robust solution for complicated and nonlinear problems due to its properties [46]. The network commonly consists of one input layer, one or more hidden layers, and one output layer. Since it does not require the assumption that data have normal distribution and linear relationships between biomass and independent variables, the ANN can deal with different data through approximation, using various complex mathematical functions, with independent variables from different data sources such as remote sensing and ancillary data [15]. A detailed overview of ANN approach is provided in Ref. 47.

Regression tree and random forest are a family of tree-based models; in the first one, data are stratified into homogeneous subsets by decreasing the within-class entropy, whereas in the second one, a large number of regression trees are constructed by selecting random bootstrap samples from the discrete or continuous dataset. In fact, the random forest algorithm is now widely used for biomass estimation [48,49].

SVM is an important method to estimate forest biophysical parameters using remote sensing data [50, 51]. It is a statistical learning algorithm with the ability to use small training sample data to produce relatively higher estimation accuracy than other approaches like ANN [15]. Ref. 51 provides a detailed overview of the SVM approach used in remote sensing. The Max Ent approach is a general-purpose machine-learning method for predicting or inferring target probability distribution from incomplete information [52]. These kinds of nonparametric algorithms have become popular in biomass modeling when large representative field datasets exist for calibration [53,54].

3. Remote sensing products for biomass modeling

3.1. From optical sources

Optical sensors are those that detect electromagnetic radiation emitted or reflected from the earth, the main source of light being the sun. Among the passive sensors are photographic and optical-electronic sensors that combine similar photographic optics and electronic detection system (detectors and push scanning) and image spectrometers [55,56]. Optical remote sensing refers to methods and technologies that acquire information from the visible, near-infrared, shortwave infrared, and thermal infrared regions of the electromagnetic spectrum. They are called optical, because energy is directed through optical components such as lenses and mirrors.

3.2. Spectral indices

Remotely sensed spectral vegetation indices represent an integrative measure of both vegetation photosynthetic activity and canopy structural variation that are widely used and have benefited numerous disciplines interested in the assessment of biomass estimation [57,58].

Different kinds of vegetation indices have been in use for a long time in AGB estimation and the number of publications is immense [6].

The key to the development of vegetation indices is the ability of the canopy of green vegetation to interact differently with certain portions of the electromagnetic spectrum. Since this contrast is particularly strong between red and near-infrared regions (NIR), it has been the focus of a large variety of attempts to build up quantitative indices of vegetation condition using remotely sensed imagery [59]. Theoretically, the ideal vegetation indices should be particularly sensitive to different vegetation covers, insensitive to soil brightness and color, and little affected by atmospheric effects [60]. However, in reality, different factors affect the reflectance of vegetation and consequently the vegetation index (e.g., atmospheric correction is essential when biomass is extracted from the vegetation indices as a final product).

According to Ref. 58, vegetation indices can be classified into (1) slope-based, (2) distance-based, and (3) transformation indices. The slope-based indices are simple arithmetic combinations that focus on the contrast between the spectral responses patterns of vegetation in the red and near-infrared portions of the electromagnetic spectrum. The most known of them are the ratio vegetation index (RVI) proposed by Birth and Mc Vey [61], normalized difference vegetation index (NDVI) introduced by Rouse et al. [62], and soil-adjusted vegetation index (SAVI) developed by Huete [63]. The vegetation indices show better sensitivity than individual spectral bands for the detection of biomass [64].

Distance-based indices measure the degree of vegetation from the soil background (known as a soil line) to the pixel with the highest content of vegetation in a perpendicular incremental distance. In this group of indices, the slope and intercept of the soil line have to be defined for each particular image; perpendicular vegetation index (PVI) introduced by Richardson and Weigand [65] cancels the effect of soil brightness in cases where vegetation is sparse and the pixels contain a mixture of green vegetation and soil background. The effect of the background soil is a major limiting factor in certain statistical analyses geared toward the quantitative assessment of AGB [59].

Transformation indices are transformations of the available spectral bands to form a new set of uncorrelated bands within which a vegetation index band can be defined. Tasseled cap (TC) is one of the most widely used indexes of this type and may apply to various remote sensing images with multiple resolutions [66–72].

In the specialized literature, many vegetation indices have been proposed, and depending on the complexity of the forest stand structure, indices vary in their relationships with biomass [23,60,73]. For forest sites with complex stand structures, vegetation indices including near-infrared wavelength have weaker relationships with biomass than those including shortwave infrared wavelength. In contrast, for areas with poor soil conditions and relatively simple forest stand structure, near-infrared vegetation indices had a strong relationship with biomass, and finally the results of transformation indices showed stronger relationships with biomass independent of different biophysical conditions [15].

3.3. Textural indices

Some studies have used regression analyses between remote sensing textural indices and biomass data from sampling sites [30,32,74–77]. This framework has been applied to different optical and synthetic aperture radar-derived indices in order to use the textural parameters as continuous spatial variables to improve biomass estimations.

In 1973, Haralick proposed a statistical analysis based on a set of parameters according to spatial dependence of gray tones in an image, defined as second-order statistics [78]. Textures are intrinsic properties of surfaces and its importance lies in image–objects segmentation, because they are related to structural arrangements of the land surface and the connections between neighboring spatial objects [79]. The most common mathematical method used to measure texture parameters is the co-occurrence matrix (gray-level co-occurrence matrix) [78]. It describes the frequency of a gray level displayed, in a specific spatial relationship, to another gray value within a neighborhood represented by a mobile window or kernel [79]. Its construction is based on four steps: (1) Window or kernel size definition, (2) band selection input, (3) texture parameters selection, and (4) spatial dependency criteria.

Textures applied parameters are described below:

Contrast	1. Dissimilarity
$$\sum_{i,j=0}^{N-1} P_{i,j}(i-j)^2$$	$$\sum_{i,j=0}^{N-1} P_{i,j}\,\lvert i-j \rvert$$
2. Homogeneity	3. Angular Second Moment
$$\sum_{i,j=0}^{N-1} \frac{P_{i,j}}{1+(i-j)^2}$$	$$\sum_{i,j=0}^{N-1} P_{i,j}^2$$
4. Maximum Probability Largest Pij value found within the window	5. Entropy $$\sum_{i,j=0}^{N-1} P_{i,j}(-\ln P_{i,j})$$
6. Mean	7. Correlation
$$\mu_i = \sum_{i,j=0}^{N-1} i(P_{i,j})$$ $$\mu_j = \sum_{i,j=0}^{N-1} j(P_{i,j})$$	$$\sum_{i,j=0}^{N-1} P_{i,j}\left[\frac{(i-\mu_i)(j-\mu_j)}{\sqrt{(\sigma_i^2)(\sigma_j^2)}} \right]$$
8. Variance	
$$\sigma_i^2 = \sum_{i,j=0}^{N-1} P_{i,j}(i-\mu_j)^2$$ $$\sigma_j^2 = \sum_{i,j=0}^{N-1} P_{i,j}(j-\mu_j)^2$$	

Table 1. Textures

$Pij = Pr$ (I$s = i \cap$ I$t = j$) = pixel probability of s is i and t is j, for separated pixels through one pixel distance in a relative displacement vector between neighboring pixels.

These texture parameters and the indices derived (e.g., ratios) have been applied in different satellite inputs, for example, on high spatial resolution optical-infrared images as SPOT-V images [30,32] and ALOS AVNIR-2 [74], medium spatial resolution resources as Landsat TM [76] and Landsat OLI [76], and SAR images as Jers-1 [76] and ALOS Palsar [32,75].

3.4. Biophysical variables

A forest biophysical parameter is a measure that simplifies the aboveground organization of plant materials [4]. In this context, several studies in remote sensing field have focused on determining canopy structural parameters such as leaf area index (LAI), canopy height or canopy fraction [80–86], the second one being the most commonly used for biomass estimation. Because canopy height is most commonly derived from active sensors [87], this part of the chapter is focused on canopy fraction.

The reflectance of forest canopy cover recorded by the instantaneous field of view (IFOV) of the sensor is a spectral mixture obtained from the interaction between electromagnetic radiation and both canopy and forest elements such as nonphotosynthetic vegetation (NPV; such as branches, stem, and litter), photosynthetic vegetation (PV; such as leaves), and others, such as bare soil and shadow. Therefore, image pixels are generally composed of more than one element, making the image interpretation difficult, which can result in a poor relation between AGB and spectral bands [88–91].

One of the most widely used remote sensing approaches to derive and extract fraction covers from mixed pixels is the spectral mixture analysis (SMA) [92]. It decomposes the mixed pixel using a collection of constituent spectra (end members) to obtain their areal proportions or abundances in a pixel, and therefore unmixing a multispectral image into fraction images of end members [88,93]; it can be linear or nonlinear. The linear model assumes a single interaction between each incident photon and the surface object, and therefore the mixed pixel is a linear combination of pure spectral signatures (end members) of the surface materials weighted by their area covered. The nonlinear model is the opposite of linear model, since electromagnetic radiation can intercept more than one element of surface, with mixed pixels resulting from a multiple-scattered signal [94].

As green leaves scatter radiation at NIR spectra, vertical structure of vegetation commonly produces a multiple-scattered signal; however, nonlinear spectral mixture approaches are barely used because they require more specific information, such as scattering properties of end members, the illumination of the sensor, and certain geometrical parameters of the scene [91,93]. For these reasons, linear approaches have been largely implemented [92].

The linear mixture model is expressed in matrix form in the following equation (3):

$$p = Mf + \varepsilon \tag{3}$$

where $p = [p1...pn]T$ is the mixed pixel, $f = [f1...fm]T$ is the fractional end-member abundance, M is an $n \times m$ matrix with n end-member spectra as column vectors, and ε is the residual not explained by the model. If p and M are known, it is possible to estimate it from ordinary least-squares procedure [93,95]), with two common constraints: the full additive condition, which determines that the fraction must sum to one, and the nonnegative condition, which makes all abundances nonnegative, thus making the model physically meaningful [92].

SMA has been successfully applied in vegetation studies to estimate the land cover fractions of PV, NPV, bare soil, and shade [96–98] or mapping the fractional cover of coniferous species [99–101]. In biomass studies, it has been proved that using ASTER fraction images (green vegetation, soil, and shade) in regression models improve the AGB estimation in Mediterranean forests, being better than NDVI or tasseled cap components [88]. Ref. [15] uses the same Landsat Thematic Mapper (TM) fractions and TM spectral signatures to relate with AGB, finding that fractional images perform better for successional forest biomass estimation than TM spectral signatures, but not for primary biomass estimation. SMA has also been applied to remove subpixel atmospheric and soil reflectance contamination in order to improve dry biomass estimations, showing that unmixed vegetation indices are better [102] than those which are not unmixed. Through geometric-optical reflectance models, Peddle et al. also estimated areal fractions of sunlit canopy, sunlit background, and shadow at subpixel scales showing higher accuracy than NDVI [103]. The other case in which sunlit crowns, background, and shadow fractions were compared with seven different vegetation indices (NDVI, SR, MSR, RDVI, WDVI, GEMI, and NLI) and three different soil-adjusted vegetation indices (SAVI, SAVI-1, and SAVI-2) to estimate biomass, LAI, net primary productivity (NPP), DBH, stem density, and basal area was the study conducted by Peddle et al. In this study, the authors concluded that the SMA shadow fraction improves the results by about 20% compared to vegetation indices [104].

3.5. From active sources

Active sensors are those that provide its own energy source in order to control the double operation of signal emission and reception of known characteristics. These sensors have the advantage of an operational capacity to produce information both at night and in the day, in addition to working in a region of the electromagnetic spectrum that makes them less sensitive to atmospheric conditions. Of these, radar and LiDAR systems [55,56] are the most known. In this section, we briefly describe each of them, pointing out the relevant examples of their application in biomass modeling.

3.6. Radio detection and ranging

RAdio detection and ranging (RADAR) is the system name of active sensors that work in microwave region of the electromagnetic spectrum. Their mechanism is performed through signal transmission–reception of a portion of energy that interacts with the surface, which is referred as backscattered, being a measure of strength and time delay of the returned signals [105].

Such energy is considered consistent or coherent (illumination beam has same wavelength and phase), which makes it possible to use different polarization schemes (orientation of emitted and detected electromagnetic fields) to generate images [106]. The spatial resolution of radar images is strongly dependent on the antenna length (aperture) of the receptor and sensor inclination angle.

Synthetic aperture radar is widely used in forest monitoring through remote sensing, which is able to generate high-resolution imagery by taking advantage of the movement of the aircraft or satellite platform. SAR simulates a long virtual antenna that comprises long coherent successive radar signals, transmitted and received by a small antenna, which simultaneously moves along a given flight path [105,106].

Since microwave energy can penetrate forest canopies, the backscattered energy of SAR systems is modulated or influenced by the structural parameters of trees (e.g., branches, leaves, and stems), which in turn depends on different ecological variables [6,107,108]. Analyses of these data have been used to determine forest state [109], forest types [110], biomass density [32,111], forest canopy height [112], forest fire degradation [49,69], deforestation [113], and forest soil moisture [114].

The sensor sensitivity to forest parameters is a function of the wavelength, for instance, bands X and C with wavelengths 2.4 and 7.5 cm, respectively, are more sensitive to backscatter of leaves, whereas bands L and P with wavelengths 15 and 100 cm, respectively, are associated with backscatter of branches and stems [55,106,115].

Three of the main approaches from SAR systems that are widely used include (1) SAR backscattering coefficient [108,111,116], (2) interferometric SAR data [32,84], and (3) polarimetric SAR data [49,117].

3.7. SAR backscattering coefficient

Forest components scatter energy transmitted by the SAR systems in all directions. A portion of energy recorded by radar is translated to a proportional ratio between density of energy scattered and density of energy transmitted from the surface targets per unit area. Backscatter coefficient ($\sigma°$) or sigma nought is the amount of radar cross section [106,108]. Generally, this magnitude is expressed as a logarithm through decibel units as the following linear-form equation (4):

$$\sigma°_{(db)} = 10*\log_{10}\sigma° \tag{4}$$

The backscatter coefficient value is related to two variables of sensor and target parameters. Sensor characteristics are a function of wavelength, polarization, and incidence angle, whereas target characteristics are associated with roughness, geometry, and dielectric properties [106, 108].

Biomass modeling through this approach has been usually applied from simple regression models under the assumption of correlation between backscatter coefficient and aboveground biomass/carbon [32,108,111,116,118]. The results are different because they rely on saturation of the signal, which is a function of wavelength, polarization, and the characteristics of the vegetation cover as well as of the difficulties caused by the specific properties of the ground as slope and aspect.

Recently, some of them have combined spatial models with remotely sensed data to improve geostatistical estimations using backscatter coefficient as spatial secondary variables [32].

3.8. Interferometry SAR data

Interferometric synthetic aperture radar (Figure 1) is a framework containing diverse methods or techniques that use phase information derived by recording phase difference or state of vibration of the wave at the instant that is received by the radar between two SAR images (known as master and slave) acquired from different sensor positions [106], called the interferometric phase ($\Delta\Phi$Int). The interferometric phase can be written as equation (5):

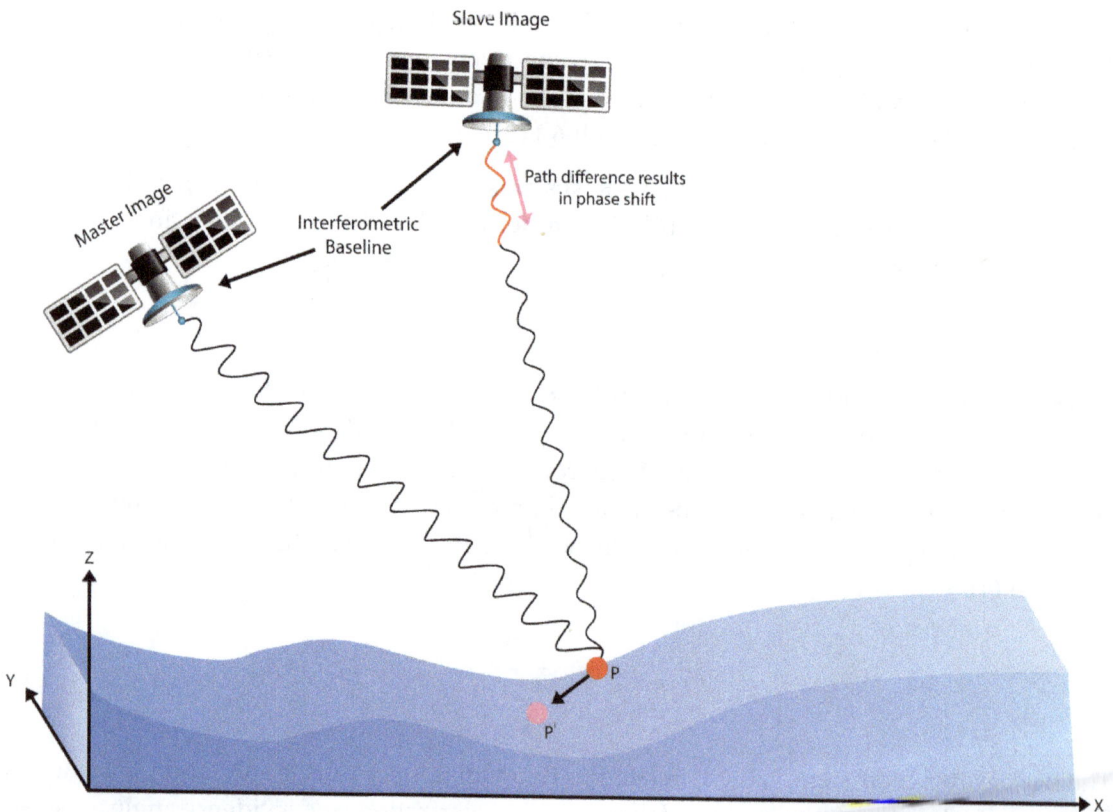

Figure 1. Interferometric synthetic aperture radar.

$$\Delta\Phi\text{INT} = \Phi S - \Phi M = \Phi\text{Topo} + \Phi\text{Mov} + \Phi\text{Atm} + \Phi\text{Noise} \tag{5}$$

where ΦS and ΦM are the phase observations of slave image and phase observations of master image, respectively, ΦTopo is the topographic component, ΦMov is the shift component, ΦAtm is the atmospheric component, and ΦNoise is the phase noise.

One parameter obtained by this approach is the coherence or correlation image as an indicator of the phase stability [105]. Its mathematical expression is (6)

$$\gamma = \frac{\left(S_1 S_2^*\right)}{\sqrt{\left(S_1 S_1^*\right)\left(S_2 S_2^*\right)}} \tag{6}$$

where γ is the coherence, the brackets $()$ denote the ensemble average, and $*$ denotes the complex conjugate; S_1 denotes the complex value of the master single-look complex (SLC) for pixels x,y and S_2^* indicates the complex conjugate value for pixels x,y in the second image known as slave. Coherence values are included between 0 and 1 and a coherence image quantifies the decorrelation between two SLC images as the loss of coherence [106]. Decorrelation is the combination of impacts in the radar phase: (1) the baseline decorrelation due to changes in the acquisition geometry of the images (which increases as the distance between the satellite orbits at the moment of acquisition increases) and (2) the temporal decorrelation due to variations in reflectivity in the zone, which can be caused by rain, phonological changes, and agricultural factors [105].

The interferometric coherence in biomass modeling is used under the assumption that for forested areas, coherence diminishes with the increment in vegetation density, as the volumetric scattering increases with movement (wind) and forest growth. Biomass modeling through this approach has been typically used from simple regression models under the assumption of correlation between interferometric coherence and aboveground biomass/carbon [32,84,116], another approach is by combining methods such as regression-kriging [32] and classification algorithms [119]. In this case, results are related to baseline and temporal decorrelations, forest type, polarization, and sensor wavelength.

3.9. Polarimetric SAR data

Antennas of radar systems can be configured to transmit and receive electromagnetic radiation polarized either horizontally or vertically. The two most common bases of polarizations are horizontal linear or H, and vertical linear or V. When the energy transmitted is polarized in the same direction as the received, it is called as like-polarized and when the transmitted energy depolarizes in a direction orthogonal to the received system, it is known as cross-polarization [120]. The polarization schemes are HH (for horizontal transmit and horizontal receive), VV (for vertical transmit and vertical receive), HV (for horizontal transmit and vertical receive), and VH (for vertical transmit and horizontal receive).

A radar system can be configured in different levels of polarization complexity:

- Single polarized – HH, VV, HV, or VH

- Dual polarized – HH and HV, VV and VH, or HH and VV

- Quad polarized – HH, VV, HV, and VH

A quadrature polarization or polarimetric radar uses these four polarizations in order to measure the magnitudes and relative phase difference between the polarization schemes or channels through an ellipse shape [106,120]. These kinds of radar systems promoted a new framework called polarimetry of synthetic aperture radar, which describes the surface through different combinations of polarization under the assumption that the interaction of electromagnetic energy and elements of the land surface can change the polarization of a portion of the wavelength transmitted by the sensor, and therefore receive information of amplitude and relative phase of the same target in four channels of information, which is considered as a basis for description of scattering polarimetric of surface targets [106,120]. It is mathematically simplified in the so-called scattering matrix (equation (7)):

$$S = \begin{bmatrix} S_{hh} & S_{hv} \\ S_{vh} & S_{vv} \end{bmatrix} \tag{7}$$

which describes different forms of the polarized electric fields between incident wave and the scatter wave in order to be the basis for diverse ways to analyze the scattering properties of a target (e.g., the covariance and coherency matrices) and diverse transformations as polarimetric decompositions and through the synergy between polarimetry and interferometry (polarimetric interferometric coherence) [120].

The use of polarimetry in biomass modeling is under the assumption of correlation between forest structural properties and polarimetric behavior. It may be construed through scattering mechanism analysis. This has been addressed mainly by polarimetric decompositions, such as Freeman Durden [49,121], eigenvector–eigenvalue [49], and Cloude and Pottier [31,121]. Biomass modeling through this framework has been usually performed from simple and multiple regression models [31,121] and nonparametric model random forest regression [49]. In this framework, results are related to forest type, spatial resolution, and sensor wavelength.
3.10. Light detection and ranging

Light detection and ranging is an active laser sensor, which emits pulses of polarized light or pulse echo, which can be calibrated within a narrow range of wavelength. The most commonly used wavelength is 1,064 nm (near-infrared), although it can range from ultraviolet to infrared range of electromagnetic spectrum (500–1500 nm) [122–124].

These laser scanners consist of a range finder, global positioning system (GPS), inertial measuring unit (IMU), and a clock capable of recording travel times to within 0.2 of a nanosecond. The integration of these systems produce accurate measurements of the position and orientation of objects registered. These technologies allow us to measure elapsed time of pulse echo between laser transmitter and objects on the surface. The energy that interacts with surfaces is backscattered over different times exhibiting multiple laser pulse returns associated with distinct surface layers toward a laser scanner that can be mounted on an aerial or terrestrial platform [123,125,126].

LiDAR information is essentially a three-dimensional point cloud composed of simple derivative returns and multiple laser pulses, this type of LiDAR data is known as discrete LiDAR returns. In addition to three-dimensional information, most LiDAR systems record the intensity as a fraction of pulse energy reflected at that location [127].

The use of LiDAR in forest areas is mainly to analyze forest vertical structural metrics under the assumption that laser can be sensed from the top of the canopies, elements of different canopies, or even to the ground, which will be reflected in the number of returns. The depth of laser penetration depends on the density of canopies and density of point clouds, which vary from less than one point per square meter to several dozens, with vertical accuracies around 12.5 cm [123,127].

One of the most widely used products in forest analysis is the result of the processing of three-dimensional point cloud, the canopy height model (CHM) (Figure 2). It is derived from the difference or subtraction between digital elevation models and digital terrain models, both datasets generally are a result of different interpolation methods, such as nearest neighbor, splines, inverse distance weighting, and kriging [127]. The first one is associated with first returns and the second one is related to the last returns. Other forest structural measurements are the fractional crown cover, crown area, crown diameter, basal area [38,125], and canopy volume [126], which are of key interest to the managers and represent information that is expensive and time consuming to collect in the field. When small-footprint LiDAR data are acquired at very high enough densities, individual tree crowns can readily be observed in the point clouds, processing algorithm for automated measuring and modeling of vegetation at individual tree crown segmentation (e.g., watershed segmentation) [95,128,129].

Biomass modeling through this approach has been usually used in simple and multiple regression models [38,125,126,130]. Other works have explored the use of learning machines [131,132]. Other approach is through combining methods that integrate LiDAR information with other sensors [22,31,124,133].

4. Case study

Mexico City is a continuous truss of multiple ecosystems, which is administratively divided into two large areas: urban land (41%) governed by the Urban Development Programs and Environmental Conservation Zone (ECZ; 61%) steered by the General Ecological Planning Program (Figure 3). The ECZ provides Mexico City with environmental services such as carbon capture, aquifer recharge, biodiversity, and scenic beauty. The zone is under anthropogenic pressure, including human settlements, land use changes, and extraction of natural resources, and therefore immediate action for conservation and appropriate resource management is necessary. This has led to deforestation, degradation, development of pest infestations, fires, and erosion. Models of the spatial variability of forest density are required in order to obtain

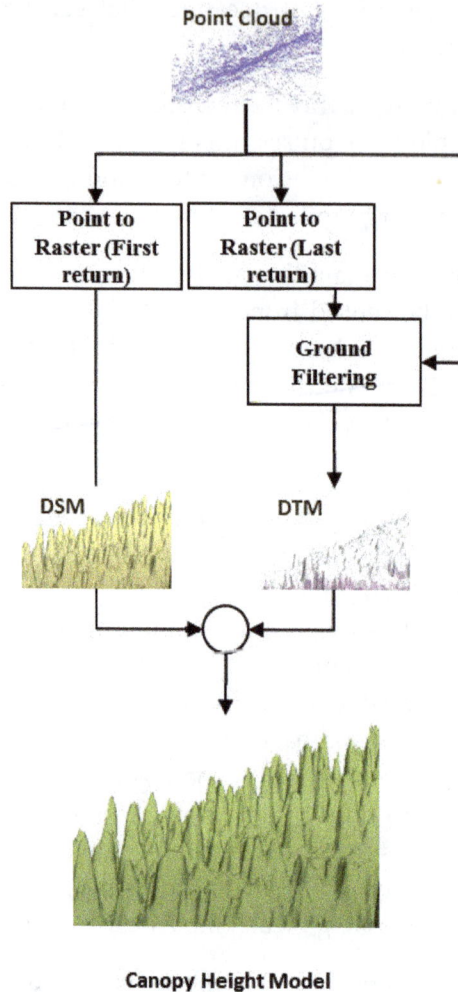

Canopy Height Model

Figure 2. Canopy height model.

an inventory of carbon storage, useful for public policies in areas with high environmental value in order to facilitate decision-making by reducing the complexity involved in integrating and interpreting values at a pixel level.

The study area lies within the ECZ of Mexico City (882 km²) and is covered by sacred fir or oyamel (*Abies religiosa*) and Mexican mountain pine (*Pinus hartwegii*) forests. Fir forests generally grow at 2,700–3,500 m above sea level. They are evergreen forests with heights of 20–40 m and their understory is densely shaded. This type of forest contains important elements, including alders (*Alnusfir mifolia*), white cedar (*Cupressus lindleyi*), oak (*Quercus laurina*), Mexican Douglas-fir (*Pseudotsuga macrolepis*), willows (*Salix oxylepis*), and black cherry (*Prunus serotina*). Pine forests (*Pinus hartwegii*) grow above 3,000 m, this species being most tolerant to the extreme environmental conditions of the high mountains. This pine develops along with Festuca and Muhlenbergia grasses [134].

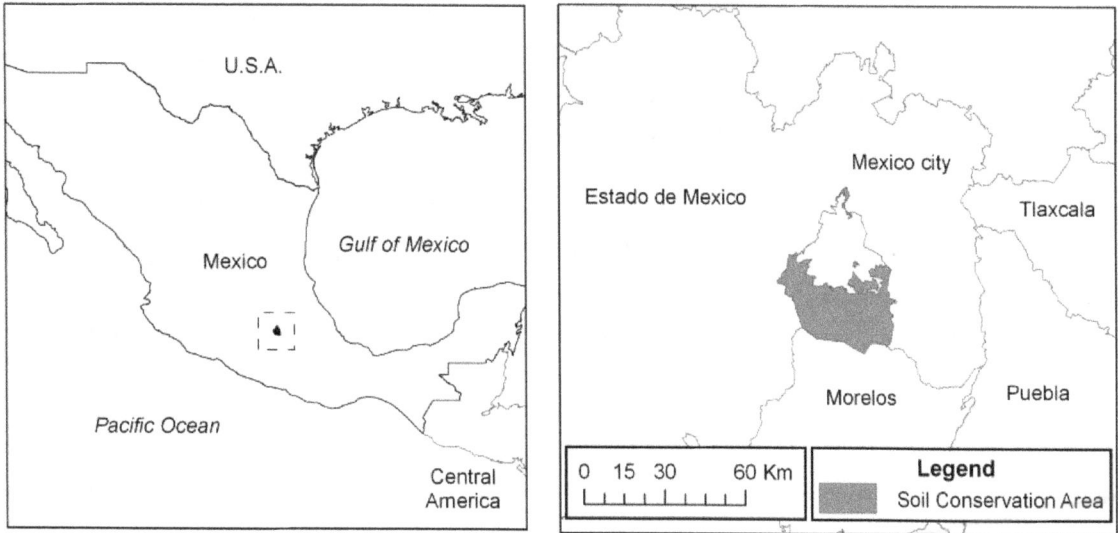

Figure 3. Environmental Conservation Zone.

The present case study is a comparative analysis between regression-cokriging and multiple regression approaches using satellite-derived indices for modeling the aboveground biomass of forests in the Environmental Conservation Zone of Mexico City. The objectives are:

1. to identify satellite-derived indices that are better associated with aboveground biomass, either from LiDAR or fraction cover (SPOT-5) imagery

2. to quantify spatial patterns of the residuals derived from simple regression between satellite indices and carbon values, using spatial autocorrelation

3. to determine whether spatial statistical methods improve the estimates of aboveground biomass carbon pools over nonspatial conventional regression methods

In order to achieve these objectives, a correlation analysis was performed between digital canopy height model (LiDAR data) and vegetation fraction cover (SPOT-5 data) and, on the other hand, ground biomass estimates at forest inventory sites. Then, the spatial autocorrelation was calculated for residuals in order to define the variables to be used in multiple models and regression-cokriging methods. Once models were obtained, the root mean square error (RMSE) was computed for each approach.

4.1. Methods

4.1.1. Models of aboveground biomass carbon used three sources of data: Forest inventory data from in situ measurements

Since aboveground carbon is the amount of carbon stored in aboveground biomass, comprising all living plant material above the soil (e.g., trunks, branches, and leaves) [3], the calculation of carbon stock from biomass consists of multiplying the total biomass by a conversion factor that represents the average carbon content in biomass. A common assumption is that biomass is around 50% carbon expressed in tons of dry matter per unit area [135]. Typically, the terms of measurement are density of biomass expressed as mass per unit area (e.g., t/ha). Here, forest inventory data were obtained from Mexico City Environmental and Land Planning Authority (PAOT, because of its name in Spanish) and were derived from sampling 283 plots during 2008–2010. Their sampling is based on the design of the National Forest and Soil Inventory of the National Forest Commission (CONAFOR, because of its name in Spanish). In it, each sampling conglomerate is composed of four circular secondary sampling plots in an inverted "Y" shape, each of which covers an area of 400 m² and peripheral plots are at 45.15 m from the center of the conglomerate (Figure 4). Of the 283 plots, 155 were among pines, 86 in fir forest, 30 in mixed forest, 10 in scrub, and 2 in planted forests. Per-tree carbon was estimated from allometric carbon equations developed by Acosta-Mireles et al., Jiménez, and Avendaño-Hernández et al. for the species of the region [136–138]. Conversion of biomass carbon from conglomerate to hectares [139] used the "ratio of means" as shown in equation (8):

$$\hat{R} = \frac{\overline{Y}}{\overline{X}} = \frac{\sum_{i=1}^{n} Y_i}{\sum_{i=1}^{n} X_i} \tag{8}$$

where Y_i is the total aboveground carbon in all plots of 400 m² and X_i is the total area sampled in i plots.

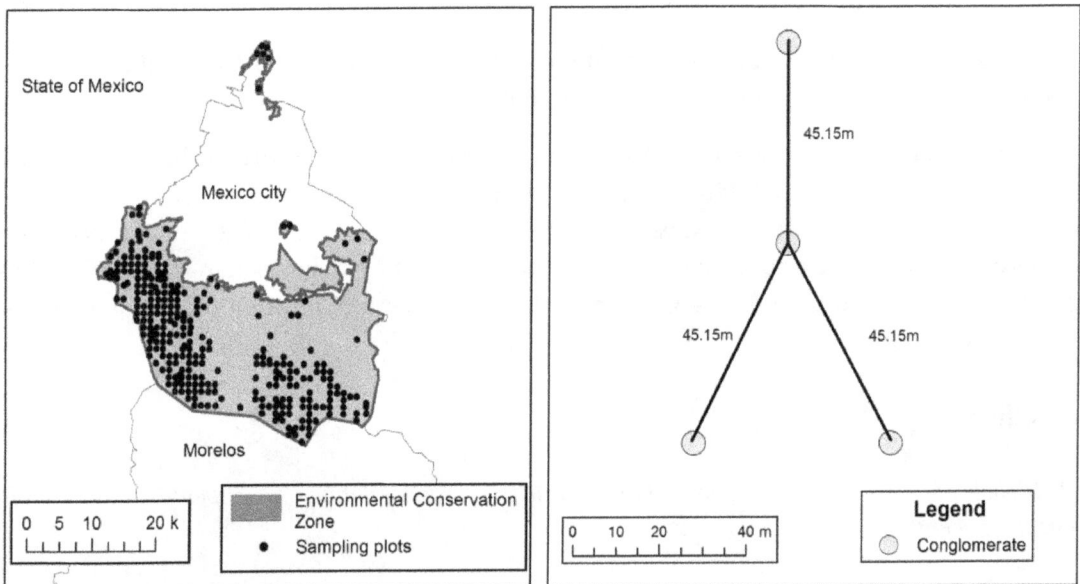

Figure 4. Sampling plots and conglomerate form.

4.1.1.1. SPOT image

Four multispectral SPOT-5 HRG images were used: two from February 25, 2010 (zenith 51.72° and 52.03°, azimuth 136.96° and 136.43°) and two from March 28, 2010 (zenith 62.67° and 62.89°, azimuth 125.66° and 124.75°). These were radiometric and atmospherically corrected with the second simulation of a satellite signal in the solar spectrum (S-6) code through CLASlite software and orthorectified with the polynomial coefficients and geoid information based on Geocover 2000 of Landsat as reference images.

4.1.2. LiDAR data

LiDAR data used were acquired from the ALS50-II sensor flown by the National Institute for Statistics and Geography (INEGI, because of its name in Spanish) between November and December of 2007 over the entire Mexico Valley. The data had an average horizontal distance of 2.0 m, minimum point density of 0.433 points/m², and vertical root mean square error of 7.3 cm. These points are used as the basis for the generation of digital terrain model (DTM) and digital surface model (DSM) with a resolution of 5 m.

4.1.2.1. Photosynthetic vegetation fraction cover

Photosynthetic vegetation fraction cover was estimated throughout the Automated Monte Carlo Unmixing (AutoMCU) model. This model integrates spectral mixture analysis and spectral end-member libraries resulting from fieldwork (ground spectrometer) and high-resolution hyperspectral information of Hyperion Sensor, in order to separate photosynthetic vegetation, non-photosynthetic vegetation, and bare substrate. The photosynthetic vegetation fraction cover was calculated by CLASlite v3.2 software (Figure 5) [140].

4.1.2.2. Canopy height model

The calculation of canopy height model used altitude values of different digital terrain model (DTM) and digital surface model (DSM) in order to extract differences between both models (Figure 6).

4.1.2.3. Correlation and autocorrelation coefficients

Correlation analysis based on multiple regressions explored statistical relationships between aboveground biomass carbon and satellite indices. The sampling points were randomly divided into 50% for model calibration and 50% for model verification. Residuals from regressions were retained and their spatial autocorrelation was quantified [141]. Moran's I index was used to identify the type and intensity of spatial pattern, measuring the degree of autocorrelation or dependence of a distribution. Moran's I index can be written as in the equation (9):

$$I = \frac{n}{\sum_i \sum_j \varpi_{ij}} * \frac{\sum_i \sum_j \varpi_{ij}\left(x_i - \bar{x}\right)\left(x_j - \bar{x}\right)}{\sum_i \left(x_i - \bar{x}\right)^2} \tag{9}$$

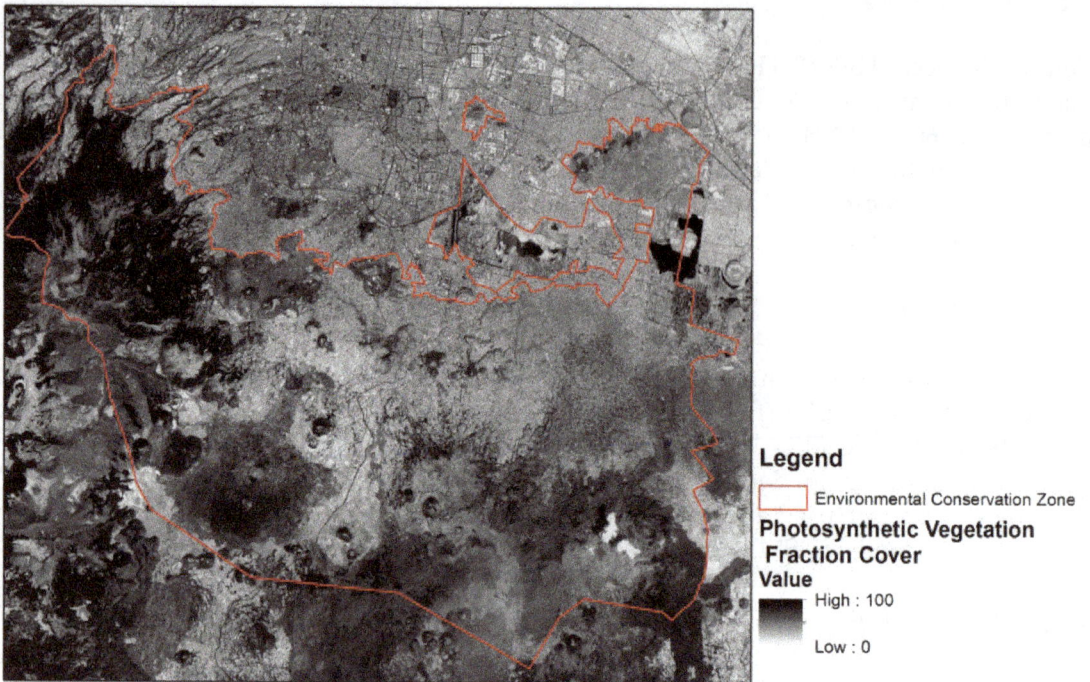

Figure 5. Photosynthetic vegetation fraction cover.

where n is the number of spatial units indexed by i and j, x is the variable of interest, and ω_{ij} is the spatial weight matrix. Moran's I values near to 1 indicate clustering negative values near to −1 represent spatial dispersion, and a value of zero indicates randomness. Statistical significance was expressed in terms of the Z descriptor and confidence level $1-\alpha$. Construction of the spatial weight matrix was distance based, since the spatial representation units are points. This distance or spatial lag includes at least 12 samples as recommended by Isaaks and Srivsatava [33].

4.1.3. Regression models

In addition to multiple regression, the present study compared models derived from regression-cokriging. The regression-cokriging was calculated through the estimation of a simple linear regression approach between aboveground carbon and canopy height model (equation (10)) and the addition of interpolated layer via ordinary cokriging integrated by regression residuals and the secondary variable (photosynthetic vegetation fraction cover) [31, 142]. The predictions were carried out separately for drift and residuals, and were added together later as in the following equation (11):

$$\hat{z}RCoK\left(S_0\right) = \hat{m}\left(S_0\right) + \hat{e}\left(S_0\right) \tag{10}$$

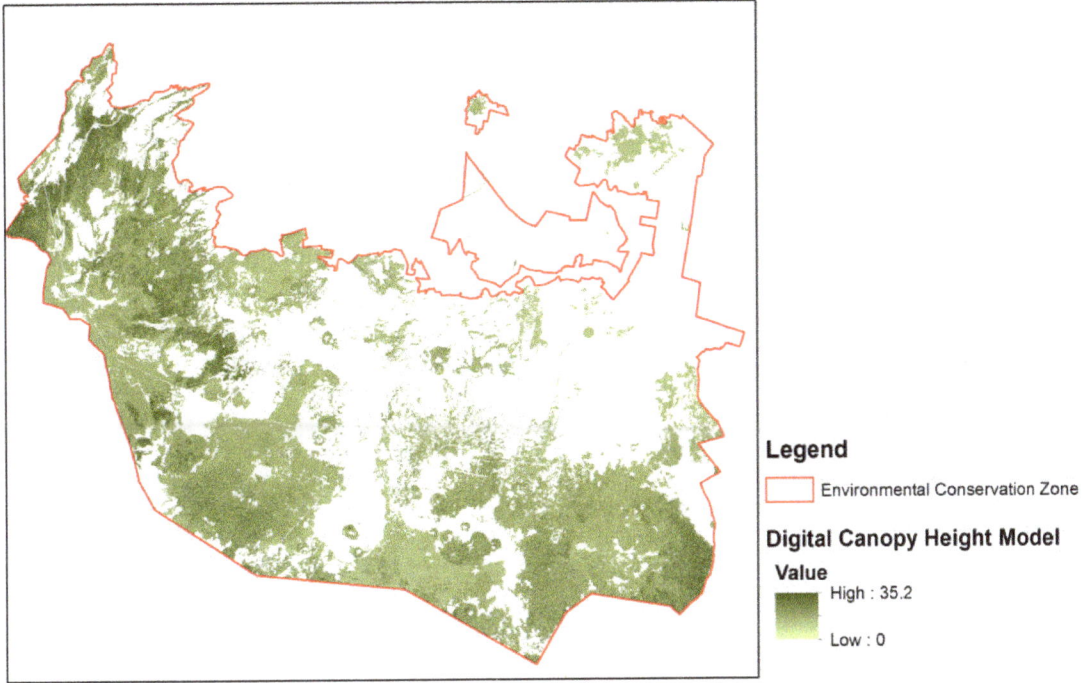

Figure 6. Digital canopy height model.

$$\hat{z}RCoK\left(S_0\right) = \sum_{k=0}^{p}\hat{\beta}_k{}^{*}q_k\left(S_0\right) + \left(\sum_{i=1}^{n}\varpi_i e\left(So\right) + \sum_{j=1}^{n}\varpi_{2j}Z_2\left(S_i\right)\right) \tag{11}$$

where βk are the coefficients of the drift model, qk is the number of auxiliary variables, $\varpi i(S_0)$ and ϖ_{2j} are the weights determined by covariogram for regression residuals, and secondary variable and e are the regression residuals [31].

4.1.3.1. Model accuracy

The regression models were validated with data from the field sampling [143]. The RMSE criterion was used to determine which regression models have more precision in the estimation of stored carbon in the area, the RMSE can be written as equation (12):

$$\text{RMSE} = \sqrt{\sum_{i=1}^{n}\frac{\left(Z_{(si)} - z_{(si)}\right)^2}{n}} \tag{12}$$

where $Z(si)$ is the reference value, $z(si)$ is the estimated value, and n is the total number of samples.

4.2. Results

4.2.1. Correlation

The degree of association between carbon stored and each index derived from remote sensing and multiple associations (Table 2) was the synergy between canopy height model and photosynthetic vegetation fraction cover with an r coefficient of 0.88. All these correlations were positive, indicating that, as stored carbon increases, vegetation indices increase.

Satellite indices	r	Error
Canopy Hieght Model	0.85	24.17
Photosynthetic Vegetation Fraction Cover	0.52	39.29
Multiple regression	0.88	22.26

Table 2. Correlation coefficients above-ground carbon and remote sensing indices.

4.2.2. Moran's I index

Once the satellite-derived index, that is the most associated with carbon storage, was identified, Moran's I was calculated from regression residuals according to sampling sites in order to identify spatial autocorrelation and, hence, the information could be included as auxiliary information in the regression-cokriging model. A value of 0.31 ($z = 2.96$, $\alpha = 0.01$) was obtained for the regression residuals between canopy height model and aboveground carbon. In this case, a low-positive spatial autocorrelation was present, with statistical significance, indicating that the neighboring spatial units presented near or close values and a slight trend toward plots.

4.2.3. Spatial distribution

To identify the spatial distribution of stored carbon, models were developed with the application of equations resulting from multiple regression and regression-cokriging spatial methods. By obtaining the covariograms for the theoretical fit, no significant variation was found between anisotropic and isotropic covariograms, therefore, the isotropic model was settled.

As the empirical covariogram showed strong spatial dependence, it did not present constant semivariances as a function of distance. In this case, the adjusted theoretical models ranged from 10,000 to 15,000 m, distances at which the observations were independent.

We used the best fit (according to root mean squares in prediction errors) for the ordinary cokriging interpolation to evaluate which is more effective in the predictions throughout the study area. Table 3 shows the parameters obtained.

Remote Sensing indices	model	Sill	Range	Nugget
Photosynthetic Vegetation Fraction Cover and regression residuals	Exponential	0.89	12,007	0

Table 3. Covariogram indices

4.2.4. Model accuracy

Comparison of the models with the set of verification sites produced the RMSE in tC/ha (Table 4). The models based on regression-cokriging presented the least error. Figure 5 shows the spatial distribution obtained by regression-cokriging and multiple regressions for stored carbon. The delineation of forest types (fir and pine) was based on the map of land use and vegetation of PAOT [144].

Models	RMSE (tC/ha)
Multiple regression	34.1
Regression-Cokriging	28.6

Table 4. RMSE

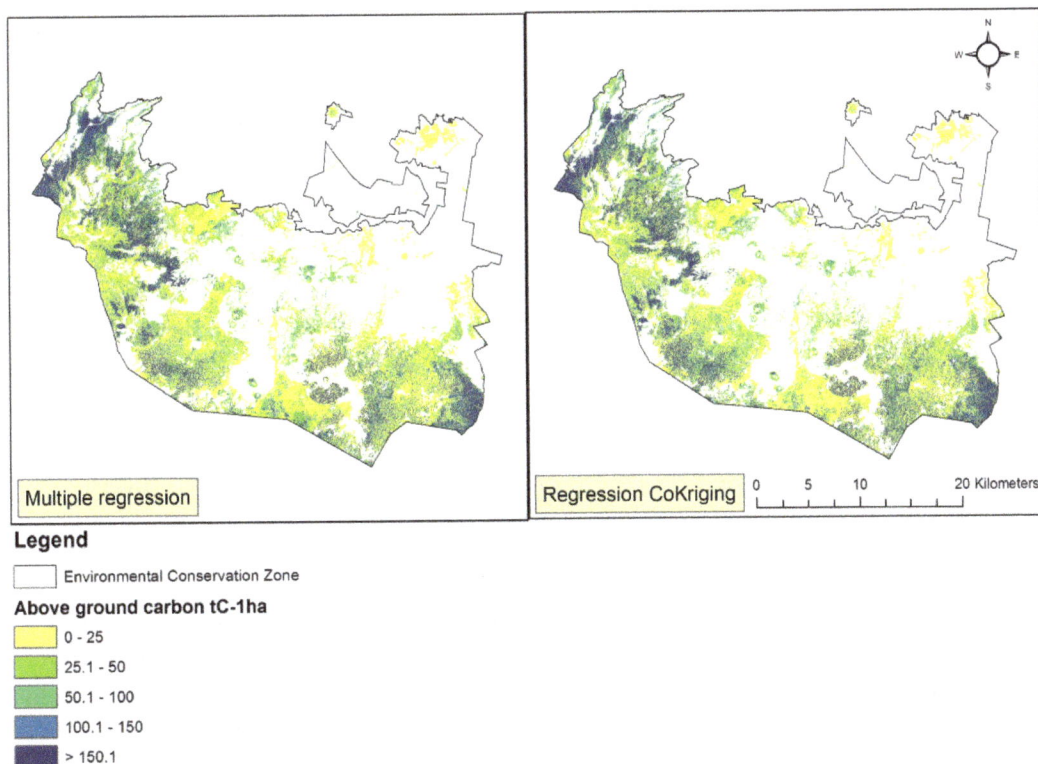

Figure 7. Spatial distribution of the aboveground carbon storage in Environmental Conservation Zone, estimated by two regression models.

The results of this research indicated that consideration of spatial autocorrelation can improve estimates of carbon content in aboveground biomass, specifically using the regression cokriging method. This could be due to its sensitivity to local variations [142], since it is particularly developed to consider the adjustment of the spatial variance model in order to improve predictions obtained from global models. The present work uses spatial modeling of indices related to carbon for the purpose of exploring the spatial autocorrelation of auxiliary variables, which is useful for representing the way in which a phenomenon radiates through spatial units. Although the primary method used for the estimation of carbon stocks in Mexico is the stratify-and-multiply approach, which assigns a single value or a range of values to each vegetation type and then multiplies these values by the areas covered by the vegetation to estimate total carbon stock values, this investigation has demonstrated a more accurate, spatial-explicit, repeatable alternative.

According to Ref. [145], autocorrelation is "perhaps, after the average and variance, the most important property of any geographic variable and, unlike these, it is explicitly linked with spatial patterns." The present comparative analysis demonstrates the importance of the use of spatial methods to model carbon stored in the aboveground biomass, since these methods consider the spatial pattern of the data. The hypothesis of the homogeneity of the relationships between stored carbon and remote sensing indices sometimes does not consider the spatial heterogeneity of many factors affecting this relationship, such as geographic differences in orientation and climatic and soil conditions [28].

This analysis provided a synoptic mapping of aboveground biomass as a potentially valuable tool for environmental protection policies in the ECZ of Mexico City, one of the most important ecological reserves for the inhabitants of the Mexico Valley in the economic, cultural, and social sense, as well as for the volume and quality of the environmental services it provides.

5. Conclusion

Remote sensing-derived indices play a major role in forest monitoring, because traditional methodologies derive their estimates of carbon content in the biomass through forest inventories and, for its implementation, they require much time and money and are generally limited to 10-year intervals. The information resulting from them is designed to present average timber volumes linked to administrative regions, which not represent the spatial variability and therefore it generate a bias in carbon measures..

This has led to great interest to estimate, map, and monitor the carbon stored in forests more precisely, enhancing the recognition of their role in the global carbon cycle, particularly in the mitigation of greenhouse gases. Through the estimation of carbon content, a base line for calculating the dynamics of this gas as a mitigation strategy can be established.

An overview of modeling options and remote sensing resources that have been used for monitoring and researching the forest biomass was presented. Techniques ranging from collecting georeferenced data in the field to the information extraction methods from satellite images and synergies between remote sensing and geostatistics were described. A case study was selected to illustrate the application of some of these techniques in the modeling of spatial distribution of aboveground carbon in Mexican coniferous forests.

Based on the study, the following conclusions can be drawn:

1. According to these results, the synergy between remote sensing and geostatistics has the potential to estimate forest biomass to improve estimations using remote sensing indices as spatial secondary variables.

2. Geospatial methods have a better modeling adjustment (e.g., RMSE) than conventional statistical methods as multiple regressions, because geospatial methods considerer local spatial variations.

3. Best Pearson coefficient between the two variables tested in the study is the digital canopy model, which resulted from LiDAR data. This kind of information is very expensive, so integrating multispectral information can be a way to capitalize on multitemporal study of biomass.

Author details

José Mauricio Galeana Pizaña*, Juan Manuel Núñez Hernández and Nirani Corona Romero

*Address all correspondence to: mgaleana@centrogeo.edu.mx

Centro de Investigación en Geografía y Geomática "Ing. Jorge L. Tamayo", México DF, Mexico

References

[1] FAO. Global Forest Resources Assessment 2010: Main Report. Roma: Food and Agriculture Organization of the United Nations. 2010. 340 p.

[2] Pavlic G, Chen W, Fernandes R, Cihlar J, Price DT, Latifovic R, Fraser R, Leblanc SG. Canada-wide maps of dominant tree species from remotely sensed and ground data. Geocarto International. 2007;22:185–204. DOI: 10.1080/10106040701201798

[3] Brown S. Measuring carbon in forests: current status and future challenges. Environmental Pollution. 2002;116:363–372. DOI: 10.1016/S0269-7491(01)00212-3

[4] Wulder M. Optical remote-sensing techniques for the assessment of forest inventory and biophysical parameters. Progress in Physical Geography. 1998;22:449–476. DOI: 10.1177/030913339802200402

[5]   Zhu X, Liu D. Improving forest aboveground biomass estimation using seasonal Landsat NDVI time-series. ISPRS Journal of Photogrammetry and Remote Sensing. 2015;102:222–231. DOI: 10.1016/j.isprsjprs.2014.08.014

[6]   Lu D. The potential and challenge of remote sensing based biomass estimation. International   Journal   of   Remote   Sensing.   2006;27:1297–1328.   DOI: 10.1080/01431160500486732

[7]   Kim Y, Yang Z, Cohen W, Pflugmacher D, Lauver C, Vankat J. Distinguishing between live and dead standing tree biomass on the North Rim of Grand Canyon National Park, USA using small-footprint lidar data. Remote Sensing of Environment. 2009;113:2499–2510. DOI: 10.1016/j.rse.2009.07.010

[8]   Bi H, Turner J, Lambert M. Additive biomass equations for native eucalypt forest trees of temperate Australia. Trees. 2004;18:467–479. DOI: 10.1007/s00468-004-0333-z

[9]   Gleason C, Jungho Im. A review of remote sensing of forest biomass and biofuel: options for small-area applications. GIScience & Remote Sensing. 2011;48:141–170. DOI: 10.2747/1548-1603.48.2.141

[10] Viana H, Aranha J, Lopes D, Cohen W. Estimation of crown biomass of *Pinus pinaster* stands and shrubland above-ground biomass using forest inventory data, remotely sensed imagery and spatial prediction models. Ecological Modelling. 2012;226:22–35. DOI: 10.1016/j.ecolmodel.2011.11.027

[11] McRoberts R, Tomppo E, Naesset E. Advances and emerging issues on national forest inventories. Scandinavian Journal of Forest Research. 2010;25:368–381. DOI: 10.1080/02827581.2010.496739

[12] Návar-Cháidez J. Biomass allometry for tree species of Northwestern Mexico. Tropical and Subtropical Agroecosystems. 2010;12:507–519.

[13] Vahedi AA, Mataji A, Babayi-Kafaki S, Eshaghi-Rad J, Hodjati SM, Djomo A. Allometric equations for predicting aboveground biomass of beech-hornbeam stands in the Hyrcanian forests of Iran. Journal of Forest Science. 2014;60:236–247.

[14] Sinha S, Jeganathan C, Sharma LK, Nathawat MS. A review of radar remote sensing for biomass estimation. International Journal of Environmental Science and Technology. 2015;12:1779–1792. DOI: 10.1007/s13762-015-0750-0

[15] Lu D, Chen Q, Wang G, Liu L, Li G, Moran E. A survey of remote sensing-based aboveground biomass estimation methods in forest ecosystems. International Journal of Digital Earth. 2014;1–43. DOI: 10.1080/17538947.2014.990526

[16] Propastin P. Modifying geographically weighted regression for estimating aboveground biomass in tropical rainforests by multispectral remote sensing data. International Journal of Applied Earth Observation and Geoinformation. 2012;18:82–90. DOI: 10.1016/j.jag.2011.12.013

[17] Kellndorfer JM, Walker WS, LaPoint E, Kirsch K, Bishop J, Fiske G. Statistical fusion of Lidar, InSAR, and optical remote sensing data for forest stand height characterization: a regional-scale method based on LVIS, SRTM, Landsat ETM+, and ancillary data sets. Journal of Geophysical Research: Biogeosciences. 2010;115:1-10. DOI: 10.1029/2009JG000997

[18] McRoberts RE, Westfall JA. Effects of uncertainty in model predictions of individual tree volume on large area volume estimates. Forest Science. 2014;60:34–42. DOI: http://dx.doi.org/10.5849/forsci.12-141

[19] Seidel D, Fleck S, Leuschner C, Hammett T. Review of ground-based methods to measure the distribution of biomass in forest canopies. Annals of Forest Science. 2011;68:225–244. DOI: 10.1007/s13595-011-0040-z

[20] Segura M, Kanninen M. Allometric models for tree volume and total aboveground biomass in a tropical humid forest in Costa Rica. Biotropica. 2005;37:2–8. DOI: 10.1111/j.1744-7429.2005.02027.x

[21] Mitchard ETA, Saatchi SS, Lewis SL, Feldpausch TR, Woodhouse IH, Sonké B, Rowland C, Meir P. Measuring biomass changes due to woody encroachment and deforestation/degradation in a forest–savanna boundary region of central Africa using multi-temporal L-band radar backscatter. Remote Sensing of Environment. 2011;115:2861–2873. DOI: 10.1016/j.rse.2010.02.022

[22] Sun G, Ranson KJ, Guo Z, Zhang Z, Montesano P, Kimes D. Forest biomass mapping from lidar and radar synergies. Remote Sensing of Environment. 2011;115:2906–2916. DOI: 10.1016/j.rse.2011.03.021

[23] Lu D. Aboveground biomass estimation using Landsat TM data in the Brazilian Amazon. International Journal of Remote Sensing. 2005;26:2509–2525. DOI: 10.1080/01431160500142145

[24] Lu D, Batistella M, Moran E. Satellite estimation of aboveground biomass and impacts of forest stand structure. Photogrammetric Engineering & Remote Sensing. 2005;71:967–974. DOI: 10.14358/PERS.71.8.967

[25] Boyd DS, Danson FM. Satellite remote sensing of forest resources: three decades of research development. Progress in Physical Geography. 2005;29:1–26. DOI: 10.1191/0309133305pp432ra

[26] Chen Q, Laurin GV, Battles JJ, Saah D. Integration of airborne lidar and vegetation types derived from aerial photography for mapping aboveground live biomass. Remote Sensing of Environment. 2012;121:108–117. DOI: 10.1016/j.rse.2012.01.021

[27] McRoberts RE, Næsset E, Gobakken T. Inference for lidar-assisted estimation of forest growing stock volume. Remote Sensing of Environment. 2013;128:268–275. DOI: 10.1016/j.rse.2012.10.007

[28] Maselli F, Chiesi M. Evaluation of statistical methods to estimate forest volume in a Mediterranean Region. IEEE Transaction on Geoscience and Remote Sensing. 2006;44:2239–2250. DOI: 10.1109/TGRS.2006.872074

[29] Sales M, Souza C, Kyriakidis P, Roberts D, Vidal E. Improving spatial distri-bution estimation of forest biomass with geostatistic: a case study for Rondonia, Brazil. Eco-logical Modelling. 2007;205:221–230. DOI: 10.1016/j.ecolmodel.2007.02.033

[30] Castillo-Santiago M, Ghilardi A, Oyama K, Hernández-Stefanoni J, Torres I, Flamen-co-Sandoval A, Fernández A, Mas J. Estimating the spatial distribution of woody bio-mass suitable for charcoal making from remote sensing and geostatistics in central Mexico. Energy for Sustainable Development. 2013;17:177–188. DOI: 10.1016/j.esd. 2012.10.007

[31] Tsui OW, Coops NC, Wulder MA, Marshall PL. Integrating airborne LiDAR and space-borne radar via multivariate kriging to estimate above-ground biomass. Re-mote Sensing of Environment. 2013;139:340–352. DOI: 10.1016/j.rse.2013.08.012

[32] Galeana-Pizaña JM, López-Caloca A, López-Quiroz P, Silván-Cárdenas J, Couturier S. Modeling the spatial distribution of above-ground carbon in Mexican coniferous forest using remote sensing and a geostatistical approach. International Journal of Applied Earth Observation and Geoinformation. 2014;30:179–189. DOI: 10.1016/j.jag. 2014.02.005

[33] Isaaks E, Srivastava R. Applied Geostatistics. New York: Oxford University Press; 1989. 561 p.

[34] Journel AG, Huijbregts ChJ. Mining Geostatistics. London: Academics Press; 1978. 600 p.

[35] Webters R, Oliver M. Geostatistics for Environmental Scientists. 2nd ed. Chichester: John Wiley & Sons; 2007. 330 p.

[36] Anselin L, Rey S, editors. Perspectives on Spatial Data Analysis. Berlin: Springer; 2010. 290 p. DOI: 10.1007/978-3-642-01976-0_1

[37] Tobler WR. A computer movie simulating urban growth in the Detroit region. Eco-nomic Geography. 1970;46:234–240. DOI: 10.2307/143141

[38] García M, Riaño D, Chuvieco E, Danson M. Estimating biomass carbon stocks for a Mediterranean forest in central Spain using LiDAR height and intensity data. Remote Sensing of Environment. 2010;114:816–830. DOI: 10.1016/j.rse.2009.11.021

[39] Mitchard ET, Saatchi SS, Baccini A, Asner GP, Goetz SJ, Harris NL, Brown S. Uncer-tainty in the spatial distribution of tropical forest biomass: a comparison of pan-trop-ical maps. Carbon Balance and Management. 2013;8:1–13. DOI: 10.1186/1750-0680-8-10

[40] McRoberts RE, GobakkenT, Næsset E. Post-stratified estimation of forest area and growing stock volume using lidar-based stratifications. Remote Sensing of Environment. 2012;125:157–166. DOI: 10.1016/j.rse.2012.07.002

[41] Saatchi S, Malhi Y, Zutta B, Buermann W, Anderson LO, Araujo AM, Phillips OL, Peacock J, ter Steege H, Lopez Gonzalez G, Baker T, Arroyo L, Almeida S, Higuchi N, Killen T, Monteagudo A, Neill D, Pitman N, Prieto A, Salomão R, Silva N, Vásquez Martínez R, Laurence W, Ramírez HA. Mapping landscape scale variations of forest structure, biomass, and productivity in Amazonia. Biogeosciences Discussions. 2009;6:5461–5505. DOI: 10.5194/bgd-6-5461-2009

[42] Kilkki P, Päivinen R. Reference sample plots to combine field measurements and satellite data in forest inventory. Department of Forest Mensuration and Management, University of Helsinki, Research Notes. 1987;19:210–215.

[43] Tomppo EO, Gagliano C, De Natale F, Katila M, McRoberts RE. Predicting categorical forest variables using an improved k-Nearest Neighbour estimator and Landsat imagery. Remote Sensing of Environment. 2009;113:500–517. DOI: 10.1016/j.rse. 2008.05.021.

[44] Latifi H, Nothdurft A, Koch B. Non-parametric prediction and mapping of standing timber volume and biomass in a temperate forest: application of multiple optical/ LiDAR-derived predictors. Forestry. 2010;83:395–407. DOI: 10.1093/forestry/cpq022

[45] Tomppo E, Olsson H, Ståhl G, Nilsson M, Hagner O, Katila M. Combining national forest inventory field plots and remote sensing data for forest databases. Remote Sensing of Environment. 2008;112:1982–1999. DOI: 10.1016/j.rse.2007.03.032

[46] Foody GM, Cutler ME, Mcmorrow J, Pelz D, Tangki H, Boyd DS, Douglas I. Mapping the biomass of Bornean tropical rain forest from remotely sensed data. Global Ecology and Biogeography. 2001;10:379–387.

[47] Mas JF, Flores JJ. The application of artificial neural networks to the analysis of remotely sensed data. International Journal of Remote Sensing. 2008;29:617–663. DOI: 10.1080/01431160701352154

[48] Pflugmacher D, Cohen WB, Kennedy RE, Yang Z. Using Landsat-derived disturbance and recovery history and lidar to map forest biomass dynamics. Remote Sensing of Environment. 2014;151:124–137. DOI: 10.1016/j.rse.2013.05.033

[49] Tanase M, Panciera R, Lowell K, Tian S, Kacker J, Walker J. Airborne multi-temporal L-band polarimetric SAR data for biomass estimation in semi-arid forests. Remote Sensing of Environment. 2014;145:93–104. DOI: 10.1016/j.rse.2014.01.024

[50] Marabel M, Alvarez-Taboada F. Spectroscopic determination of aboveground biomass in grasslands using spectral transformations, support vector machine and partial least squares regression. Sensors. 2013;13:10027–10051. DOI: 10.3390/s130810027

[51] Mountrakis G, Im J, Ogole C. Support vector machines in remote sensing: a review. ISPRS Journal of Photogrammetry and Remote Sensing. 2011;66:247–259. DOI: 10.1016/j.isprsjprs.2010.11.001

[52] Phillips SJ, Dudík M. Modeling of species distributions with Maxent: new extensions and a comprehensive evaluation. Ecography. 2008;31:161–175. DOI: 10.1111/j. 0906-7590.2008.5203.x

[53] Harris NL, Brown S, Hagen SC, Saatchi SS, Petrova S, Salas W, Hansen MC, Potapov PV, Lotsch A. Baseline map of carbon emissions from deforestation in tropical regions. Science. 2012;336:1573–1576. DOI: 10.1126/science.1217962

[54] Saatchi SS, Harris NL, Brown S, Lefsky M, Mitchard ET, Salas W, Zutta BR, Buermann W, Lewis SL, Hagen S, Petrova S, White L, Silman M, Morel A. Benchmark map of forest carbon stocks in tropical regions across three continents. Proceedings of the National Academy of Sciences of the United States of America. 2011;108:9899–9904. DOI: 10.1073/pnas.1019576108

[55] Franklin S. Remote sensing for sustainable forest management. USA: CRC Press LLC; 2001. 409 p.

[56] Jensen, J. Remote Sensing of the Environment: An Earth Resource Perspective. 2nd ed. New Jersey: Prentice Hall; 2006. 608 p.

[57] Mundava C, Helmholz P, Schut T, Corner R, McAtee B, Lamb D. Evaluation of vegetation indices for rangeland biomass estimation in the Kimberley area of Western Australia. ISPRS Journal of Photogrammetry and Remote Sensing. 2014;2:47–53. DOI: 10.5194/isprsannals-II-7-47-2014

[58] Jackson RD, Huete AR. Interpreting vegetation indices. Preventive Veterinary Medicine. 1991;11:185–200. DOI: 10.1016/S0167-5877(05)80004-2

[59] Silleos NG, Alexandridis TK, Gitas IZ, Perakis K. Vegetation indices: advances made in biomass estimation and vegetation monitoring in the last 30 years. Geocarto International. 2006;21:21–28. DOI: 10.1080/10106040608542399

[60] Bannari A, Morin D, Bonn F, Huete AR. A review of vegetation indices. Remote Sensing Reviews. 1995;13:95–120. DOI: 10.1080/02757259509532298

[61] Birth GS, McVey GR. Measuring the color of growing turf with a reflectance spectrophotometer. Agronomy Journal. 1986;60:640–643. DOI: 10.2134/ agronj1968.00021962006000060016x

[62] Rouse JW, Haas RH, Deering DW, Harlan JC. Monitoring the vernal advancement and retrogradation (green wave effect) of natural vegetation (Technical report). Texas: A & M University, Remote Sensing Center. 1974. 8 p.

[63] Huete AR. A soil-adjusted vegetation index (SAVI). Remote Sensing of Environment. 1988;25:295–309. DOI: 10.1016/0034-4257(88)90106-X

[64] Asrar G, Fuchs M, Kanemasu ET, Hatfield JL. Estimating absorbed photosynthetic radiation and leaf area index from spectral reflectance in wheat. Agronomy Journal. 1984;76:300–306. DOI: 10.2134/agronj1984.00021962007600020029x

[65] Richardson AJ, Weigand CL. Distinguishing vegetation from soil background information. Photogrammetric Engineering and Remote Sensing. 1977;43:1541–1552.

[66] Baig MHA, Zhang L, Shuai T, Tong Q. Derivation of a tasselled cap transformation based on Landsat 8 at-satellite reflectance. Remote Sensing Letters. 2014;5:423–431. DOI: 10.1080/2150704X.2014.915434

[67] Yarbrough LD, Easson G, Kuszmaul JS. Using at-sensor radiance and reflectance tasseled cap transforms applied to change detection for the ASTER sensor. ASTER. 2005;2:141–145.

[68] Horne JH. A tasseled cap transformation for IKONOS images. In: Proceedings of the InASPRS 2003 Annual Conference, May 2003; Anchorage Alaska; 2003. pp. 60–70.

[69] Huang C, Wylie B, Yang L, Homer C, Zylstra G. Derivation of a tasselled cap transformation based on Landsat 7 at-satellite reflectance. International Journal of Remote Sensing. 2002;23:1741–1748. DOI: 10.1080/01431160110106113

[70] Zhang X, Schaaf CB, Friedl M, Strahler AH, Gao F, Hodges JC. MODIS tasseled cap transformation and its utility. In: Proceedings of the Geoscience and Remote Sensing Symposium, IGARSS'02. 2002 IEEE International (Vol. 2); 24–28 June 2002. Canada: IEEE; 2002. pp. 1063–1065.

[71] Crist EP. A TM tasseled cap equivalent transformation for reflectance factor data. Remote Sensing of Environment. 1985;17:301–306. DOI: 10.1016/0034-4257(85)90102-6

[72] Kauth RJ, Thomas GS. The tasseled cap — A graphic description of the spectral-temporal development of agricultural crops as seen by Landsat. In: Proceedings of the Symposium on Machine Processing of Remotely Sensed Data; June 29–July 1 1976; West Lafayette Indiana. 1976. pp. 41–51.

[73] McDonald AJ, Gemmell FM, Lewis PE. Investigation of the utility of spectral vegetation indices for determining information on coniferous forests. Remote Sensing of Environment. 1998;66:250–272. DOI: 10.1016/S0034-4257(98)00057-1

[74] Sarker L, Nichol J. Improved forest biomass estimates using ALOS AVNIR-2 texture indices. Remote Sensing of Environment. 2011;115:968–977. DOI: 10.1016/j.rse.2010.11.010

[75] Sarker L, Nichol J, Ahmad B, Busu I, Rahman AA. Potential of texture measurements of two-date dual polarization PALSAR data for the improvement of forest biomass estimation. ISPRS Journal of Photogrammetry and Remote Sensing. 2012;69:146–166. DOI: 10.1016/j.isprsjprs.2012.03.002

[76] Cutler MEJ, Boyd DS, Foody GM, Vetrivel A. Estimating tropical forest biomass with a combination of SAR image texture and Landsat TM data: an assessment of predic-

tions between regions. ISPRS Journal of Photogrammetry and Remote Sensing. 2012;70:66–77. DOI: 10.1016/j.isprsjprs.2012.03.011

[77] Dube T, Mutanga O. Investigating the robustness of the new Landsat-8 Operational Land Imager derived texture metrics in estimating plantation forest aboveground biomass in resource constrained areas. ISPRS Journal of Photogrammetry and Remote Sensing. 2015;108:12–32. DOI: 10.1016/j.isprsjprs.2015.06.002

[78] Haralick R, Shanmugam K, Dinstein I. Textural features for image classification. IEEE Transaction on Systems, Man, and Cybernetics. 1973;6:610–621. DOI: 10.1109/TSMC.1973.4309314

[79] Richards JA. Remote Sensing Digital Image Analysis (Vol. 3). Berlin: Springer; 1999

[80] Ozdemir I, Donoghue D. Modelling tree size diversity from airborne laser scanning using canopy height model with image textures measures. Forest Ecology and Management. 2013;295:28–37. DOI: 10.1016/j.foreco.2012.12.044

[81] Sharma R, Kajiwara K, Honda Y. Estimation of forest canopy structural parameters using kernel-driven bi-directional reflectance model based multi-angular vegetation indices. ISPRS Journal of Photogrammetry and Remote Sensing. 2013;78:50–57. DOI: 10.1016/j.isprsjprs.2012.12.006

[82] Mura N, Jones D. Characterizing forest ecological structure using pulse types and heights of airborne laser scanning. Remote Sensing of Environment. 2010;114:1069–1076. DOI: 10.1016/j.rse.2009.12.017

[83] Mandugundu R, Nizalapur V, Jha C. Estimation of LAI and above-ground biomass in deciduous forest: Western Ghats of Karnataka, India. International Journal of Applied Earth Observation and Geoinformation. 2008;10:211–219. DOI: 10.1016/j.jag.2007.11.004

[84] Solberg S, Astrup R, Gobakken T, Næsset E, Weydahl D. Estimating spruce and pine biomass with interferometric X-band SAR. Remote Sensing of Environment. 2010;114:2353–2360. DOI: 10.1016/j.rse.2010.05.011

[85] Heiskanen J. Estimating aboveground tree biomass and leaf area index in mountain birch forest using ASTER satellite data. International Journal of Remote Sensing. 2006;27:1135–1158. DOI: 10.1080/01431160500353858

[86] White J, Running S, Nemani R, Keane R, Ryan K. Measurement and remote sensing of LAI in Rocky Mountain montane ecosystems. Canadian Journal of Forest Research. 1997;27:1714–1727.

[87] Donoghue D, Watt P, Cox N, Wilson J. Remote sensing of species mixtures in conifer plantations using LIDAR height and intensity data. Remote Sensing of Environment. 2007;110:509–522. DOI: 10.1016/j.rse.2007.02.032

[88] Fernández-Manso O, Fernández-Manso A, Quintano C. Estimation of aboveground biomass in Mediterranean forests by statistical modelling of ASTER fraction images.

International Journal of Applied Earth Observation and Geoinformation. 2014;31:45–56. DOI: 10.1016/j.jag.2014.03.005

[89] Cabacinha CD, de Castro SS. Relationship between floristic diversity and vegetation indices, forest structure and landscape metrics of fragments in Brazilian Cerrado. Forest Ecology and Management. 2009;257:2157–2165. DOI: 10.1016/j.foreco. 2009.02.030

[90] Gilabert MAJ, González-Piqueras J, García-Haro J. Acerca de los índices de vegetación. Revista de Teledetección. 1997;8:1–10.

[91] Roberts DA, Smith MO, Adams JB. Green vegetation, nonphotosynthetic vegetation and soils in AVIRIS data. Remote Sensing of Environment. 1993;44:255–44:255–44:255–44:255–44:255–44:255–269

[92] Quintano C, Fernández-Manso A, Shimabukuro YE, Pereira G. Spectral unmixing. International Journal of Remote Sensing. 2012;33:5307–5340. DOI: 10.1080/01431161.2012.661095

[93] Keshava N, Mustard JF. Spectral unmixing. IEEE. Signal Processing Magazine. 2002; 19:44–57.

[94] Somers B, Asner GP, Tits L, Coppin P. Endmember variability in spectral mixture analysis: a review. Remote Sensing of Environment. 2011;115:1603–1616. DOI: 10.1016/j.rse.2011.03.003

[95] Silván-Cárdenas JL, Wang L. Retrieval of subpixel Tamarix canopy cover from Landsat data along the Forgotten River using linear and nonlinear mixture models. Remote Sensing of Environment. 2010; 114:1777-1790. DOI: 10.1016/j.rse.2010.04.003

[96] Okin GS, Clarke KD, Lewis MM. Comparison of methods for estimation of absolute vegetation and soil fractional cover using MODIS normalized BRDF-adjusted reflectance data. Remote Sensing of Environment. 2013;130:266–279. DOI: 10.1016/j.rse. 2012.11.021

[97] Yang J, Weisberg PJ, Bristow NA. Landsat remote sensing approaches for monitoring long-term tree cover dynamics in semi-arid woodlands: comparison of vegetation indices and spectral mixture analysis. Remote Sensing of Environment. 2012;119:62–71. DOI: 10.1016/j.rse.2011.12.004

[98] Souza Jr. C, Firestone L, Moreira Silva L, Roberts D. Mapping forest degradation in the Eastern Amazon from SPOT 4 through spectral mixture models. Remote Sensing of Environment. 2003;87:494–506. DOI: 10.1016/j.rse.2002.08.002

[99] Olthof I, Fraser RH. Mapping northern land cover fractions using Landsat ETM+. Remote Sensing of Environment. 2007;107:496–509. DOI: 10.1016/j.rse.2006.10.009

[100] Vikhamar D, Solberg R. Subpixel mapping of snow cover in forests by optical remote sensing. Remote Sensing of Environment. 2002;84:69–82. DOI: 10.1016/S0034-4257(02)00098-6

[101] Lobell DB, Asner GP, Law BE, Treuhaft RN. View angle effects on canopy reflectance and spectral mixture analysis of coniferous forest using AVIRIS. International Journal of Remote Sensing. 2002;23:2247–2262. DOI: 10.1080/01431160110075613

[102] Huang J, Chen D, Cosh M. Sub-pixel reflectance unmixing in estimating vegetation water content and dry biomass of corn and soybeans cropland using normalized difference water index (NDWI) from satellites. Remote Sensing of Environment. 2009;30:2075–2104. DOI: 10.1080/01431160802549245

[103] Peddle DR, Hall FG, LeDrew EF. Spectral mixture analysis and geometric-optical reflectance modeling of boreal forest biophysical structure. Remote Sensing of Environment. 1999;67:288–297. DOI: 10.1016/S0034-4257(98)00090-X

[104] Peddle DR, Brunke SP, Hall FG. A comparision of spectral mixture analysis and ten vegetation indices for estimating boreal forest biophysical information from airborne data. Canadian Journal of Remote Sensing: Journal canadien de télédétection. 2001;27:627–635. DOI: 10.1080/07038992.2001.10854903

[105] Maître H, editor. Processing of Synthetic Aperture Radar Images. Hoboken: John Wiley & Sons, Inc; 2008. 384 p.

[106] Massonnet D, Souyris JC. Imaging with Synthetic Aperture Radar. Boca Raton: CRC Press; 2008. 250 p.

[107] Goetz S, Baccini A, Laporte N, Johns T, Walker W, Kellndorfer J, Houghton R, Sun M. Mapping and monitoring carbon stocks with satellite observations: a comparison of methods. Carbon Balance and Management. 2009;4:1–7. DOI: 10.1186/1750-0680-4-2

[108] Collins J, Hutley L, Williams R, Boggs G, Bell D, Bartolo R. Estimating landscape-scale vegetation carbon stocks using airborne multi-frequency polarimetric synthetic aperture radar (SAR) in the savannahs of north Australia. International Journal of Remote Sensing. 2009;30:1141–1159. DOI: 10.1080/01431160802448935

[109] Thirion-Leferevre L, Colin-Koeniguer E. Investigating attenuation, scattering phase center and total height using simulated interferometric SAR images of forested areas. IEEE Transaction on Geoscience and Remote Sensing. 2007;45:3172–3179. DOI: 10.1109/TGRS.2007.904921

[110] Laurin G V, Liesenberg V, Chen Q, Guerriero L, Del Frate F, Bartolini A, Valentini R. Optical and SAR sensor synergies for forest and land cover mapping in a tropical site in West Africa. International Journal of Applied Earth Observation and Geoinformation. 2013;21:7–16. DOI: 10.1016/j.jag.2012.08.002

[111] Cartus O, Santoro M, Kellndorfer J. Mapping forest aboveground biomass in the Northeastern United States with ALOS PALSAR dual-polarization L-band. Remote Sensing of Environment. 2012;124:466–478. DOI: 10.1016/j.rse.2012.05.029

[112] Karjalainen M, Kankare V, Vastaranta M, Holopainen M, Hyyppä J. Prediction of plot-level forest variables using TerraSAR-X stereo SAR data. Remote Sensing of Environment. 2012;117: 338–347. DOI: 10.1016/j.rse.2011.10.008

[113] Schlund M, von Poncet F, Hoekman DH, Kuntz S, Schmullius C. Importance of bistatic SAR features from TanDEM-X for forest mapping and monitoring. Remote Sensing of Environment. 2014;151:16–26. DOI: 10.1016/j.rse.2013.08.024

[114] Bourgeau-Chavez L, Leblon B, Charbonneau F, Buckley J. Evaluation of polarimetric Radarsat-2 SAR data for development of soil moisture retrieval algorithms over a chronosequence of black spruce boreal forest. Remote Sensing of Environment. 2013;132:71–85. DOI: 10.1016/j.rse.2013.01.006

[115] Balzter H. Forest mapping and monitoring with interferometric Synthetic Aperture Radar (InSAR). Progress in Physical Geography. 2001;25:159–177. DOI: 10.1177/030913330102500201

[116] Santoro M, Fransson J, Eriksson L, Magnusson M, Ulander L, Olsson H. Signatures of ALOS PALSAR L-band backscatter in Swedish forest. IEEE Transaction on Geoscience and Remote Sensing. 2009;47:4001–4019. DOI: 10.1109/TGRS.2009.2023906

[117] Le Toan T, Quegan S, Davidson M, Balzter H, Paillou P, Papathanassiou Plummer S, Rocca F, Saatchi S, Shugart H, Ulander L. The BIOMASS mission: mapping global forest biomass to better understand the terrestrial carbon cycle. Remote Sensing of Environment. 2011;115:2850–2860. DOI: 10.1016/j.rse.2011.03.020

[118] Carreiras J, Vasconcelos M, Lucas R. Understanding the relationship between aboveground biomass and ALOS PALSAR data in the forest of Guinea-Bissau (West Africa). Remote Sensing of Environment. 2012;121:426–442. DOI: 10.1016/j.rse.2012.02.012

[119] Tansey KJ, Luckman AJ, Skinner L, Balzter H, Strozzi T, Wagner W. Classification of forest volume resources using ERS tandem coherence and JERS backscatter data. International Journal of Remote Sensing. 2004;25:751–768. DOI: 10.1080/0143116031000149970

[120] Lee JS, Pottier E. Polarimetric radar imaging: from basics to applications. Boca Raton: CRC press; 2009. 391 p. DOI: 10.1201/9781420054989.fmatt

[121] Gonçalves F, Santos J, Treuhaft R. Stem volume of tropical forest from polarimetric radar. International Journal of Remote Sensing. 2011;32:503–522. DOI: 10.1080/01431160903475217

[122] Lefsky MA, Cohen W, Parker G, Harding D. LiDAR remote sensing for ecosystem studies. BioScience. 2002;52:19–30.

[123] Andersen H, Reutebuch S, McGaughey R. A rigorous assessment of tree height measurements obtained using airborne LiDAR and conventional field methods. Canadian Journal of Remote Sensing. 2006;32:355–366.

[124] Hyde P, Dubayah R, Walker W, Blair JB, Hofton M, Hunsaker C. Mapping forest structure for wildlife habitat analysis using multi-sensor (LiDAR, SAR/InSAR, ETM+, Quickbird) synergy. Remote Sensing of Environment. 2006;102:63–73. DOI: 10.1016/j.rse.2006.01.021

[125] Chopping M, Nolin A, Moisen G, Martonchik J, Bull M. Forest canopy height from the multiangle imaging spectroradiometer (MISR) assessed with high resolution discrete return lidar. Remote Sensing of Environment. 2009;113:2172–2185. DOI: 10.1016/j.rse.2009.05.017

[126] Vauhkonen J, Næsset E, Gobakken T. Deriving airborne laser scanning based computational canopy volume forest biomass and allometry studies. ISPRS Journal of Photogrammetry and Remote Sensing. 2014;96:57–66. DOI: 10.1016/j.isprsjprs.2014.07.001

[127] Hyyppä J, Hyyppä H, Yu X, Kaartinen H, Kukko A, Holopainen M. Forest inventory using small-footprint airborne LiDAR. Finland. In: Shan J, Toth CK, editors. Topographic Laser Ranging and Scanning: Principles and Processing. CRC Press; 2008. pp. 335–370. DOI: 10.1201/9781420051438.ch12

[128] Chen Q, Baldocchi D, Gong P, Kelly M. Isolating individual trees in a savanna woodland using small footprint lidar data. Photogrammetric Engineering and Remote Sensing. 2006;72:923–932. DOI: 0099-1112/06/7208–0923/$3.00/0

[129] Hyyppa J, Kelle O, Lehikoinen M, Inkinen M. A segmentation-based method to retrieve stem volume estimates from 3-D tree height models produced by laser scanners. IEEE Transactions on Geoscience and Remote Sensing. 2001;39:969–975. DOI: 10.1109/36.921414

[130] McRoberts R, Næsset E, Gobakken T, Martin O. Indirect and direct estimation of forest biomass change using forest inventory and airborne laser scanning data. Remote Sensing of Environment. 2015;164:36–42. DOI: 10.1016/j.rse.2015.02.018

[131] Gleason C, Jungho Im. Forest biomass estimation from airborne LiDAR data using machine learning approaches. Remote Sensing of Environment. 2012;125:80–91. DOI: 10.1016/j.rse.2012.07.006

[132] García-Gutiérrez J, Martínez Álvarez F, Troncoso A, Riquelme J. A comparison of machine learning regression techniques for LiDAR-derived estimation of forest variables. Neurocomputing. 2015;167:24–31. DOI: 10.1016/j.neucom.2014.09.091

[133] Clark M, Roberts D, Ewel J, Clark D. Estimation of tropical rain forest aboveground biomass with small-foorprint lidar and hyperspectral sensors. Remote Sensing of Environment. 2011;115:2931–2942. DOI: 10.1016/j.rse.2010.08.029

[134] Rzedowski J. Vegetación de México. Mexico: Limusa; 1978. 432 p.

[135] Houghton R. Aboveground forest biomass and the global carbon balance. Global Change Biology. 2005;11:945–958. DOI: 10.1111/j.1365-2486.2005.00955.x

[136] Acosta-Mireles M, Vargas-Hernández J, Velásquez-Martínez A, Etchevers-Barra JD. Estimación de la biomasa aérea mediante el uso de relaciones alométricas en seis especies arbóreas en Oaxaca, México. Agrociencia. 2002;36:725–736.

[137] Jiménez M. Ecuaciones alométricas para determinar biomasa y carbono para *Pinus hartwegii* en el Parque Nacional Izta- Popo [thesis]. Texcoco: Universidad Autónoma Chapingo; 2009.

[138] Avendaño-Hernández DM, Acosta-Mireles M, Carrillo-Anzures F, Etchevers-Barra JD. Estimación de biomasa y carbono en un bosque de Abies religiosa. Fitotecnia Mexicana. 2009;32:233–238.

[139] Šmelko Š, Merganič J. Some methodological aspects of the National Forest Inventory and Monitoring in Slovakia. Journal of Forest Science. 2008;54:476–483.

[140] Asner GP, Heidebrecht KB. Spectral unmixing of vegetation, soil and dry carbon cover in arid regions: comparing multispectral and hyperspectral observations. International Journal of Remote Sensing. 2002;23:3939–3958. DOI: 10.1080/01431160110115960

[141] Anselin L, Rey S, editors. Perspectives on Spatial Data Analysis. New York: Springer; 2010. 290 p. DOI: 10.1007/978-3-642-01976-0

[142] Hengl T, Heuvelink GBM, Stein A, 2003. Comparison of kriging with external drift and regression-kriging. Technical note, ITC, WEB. 3 august 2015. Available online at https://www.itc.nl/library/Papers_2003/misca/hengl_comparison.pdf

[143] Goovaerts P. Geostatistical approaches for incorporating elevation into the spatial interpolation of rainfall. Journal of Hydrology. 2000;228:113–129. DOI: 10.1016/S0022-1694(00)00144-X

[144] GDF. Atlas geográfico del suelo de conservación del Distrito Federal. Ciudad de México: Gobierno del Distrito Federal, Secretaría del Medio Ambiente, Procuraduría Ambiental y del Ordenamiento Territorial del Distrito Federal; 2012. 96 p.

[145] Celemín JP. Autocorrelación espacial e indicadores locales de asociación espacial: Importancia, estructura y aplicación. Revista Universitaria de Geografía. 2009;18:11–31.

6

Climate Factors' Effects on Glacier Variations in the Commune of Alto del Carmen, Chile

Guido Staub and Catherinne Muñoz

Additional information is available at the end of the chapter

Abstract

Ice bodies in the semi-arid mountainous regions of Chile are of vital importance for the local population. As variations of their extent are often associated with climate change, this study focuses on the glaciers and glacierets situated in the Commune Alto del Carmen and local and regional climate. We combine statistically Landsat satellite imagery, historical and ongoing weather data. The present study covers a time span of 21 years, 1994–2015. Our results indicate that the extent of all ice bodies has continuously diminished as a consequence of long-term climate variability.

Keywords: Glaciers, Andes, Climate ENSO

1. Introduction

Since the end of the Little Ice Age, from about 1300 to about 1850, many worldwide glaciers have decreased in volume and extent [1]. Therefore, some of them at present are finished and others will disappear in the near future [2, 3]. It is understood, but still not very well documented, that glacier retreat is closely coupled to global climate change and anthropogenic interventions [4, 5, 6, 7]. This is mainly due to the complexity of weather system because of difficult climate history reconstruction. The main causes of receding glaciers, which can be attributed to climate variations, are constantly increasing global and regional temperature and lower stationary precipitation in the affected areas. Glaciologists have found out that the phenomenon of glacier retreat coincides with an increase in greenhouse gas emissions during and after the industrial revolution in the 18th century, see Figure 1. This means that human activities play a major role in this context. Even in a more direct way humans intervene, as exploration and exploitation of nature are activities that can be dated back till the beginning of the new age. It is quite clear that nowadays these interventions are carried out in a different way due to the possibility to use highly developed machines, and that the principal purpose has changed as in the past natural resources were exploited by humans as personal necessities had to be covered. At present, more economic reasons are in the foreground [8, 9, 10].

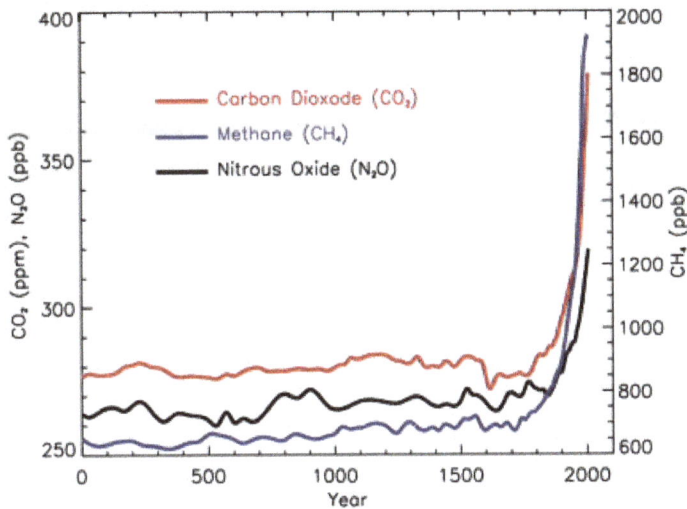

Figure 1. Trend of greenhouse gas emissions over the past 2,000 years (Source: [11])

Nevertheless, by all the interventions that are carried out by humans, ecosystems, our planet in general, suffer continuous alterations. But it would be simple to attribute every trend to humans as there are also natural circumstances, seasonal, periodic and single events, which affect nature. In any case, regardless of what or who have responsibility, the phenomenon of glacier retreat has to be studied as its impact is huge and might affect a whole region or even a country [12].

It has to be mentioned that glacier retreat should not be confused with other cyclical phenomena, like some melt during spring and summer months, which have almost no negative impact on glaciers. Annual thaw starts each spring in the mountains and causes melting of snow and ice accumulated during winter. The melting snow during spring and summer months causes an overall positive impact, since it generates a valuable source of fresh water. During winter months, snow fall results in a recuperation of melted snow. In consequence, an almost neutral mass balance between warm and cold periods is achieved by nature as this process repeats year after year. This is not the case if glaciers melt as there is a negative mass balance during a certain period of time. So the problem arises when the phenomenon is not seasonal, the glacier does not recover its initial volume in the cold months, year after year, so its volume and extend gets diminished and in consequence, natural fresh water source for human consumption and irritation is threatened [13].

In South America, a total surface of about 26,000 km² is covered by glaciers. Almost 77% of this area can be found in Chile [14]; considering 1,751 glaciers (16,893 sqkm) already mapped and 5,000 sqkm estimated of not yet registered glaciers. Their distribution from north to south can be grouped as it is shown in Table 1.

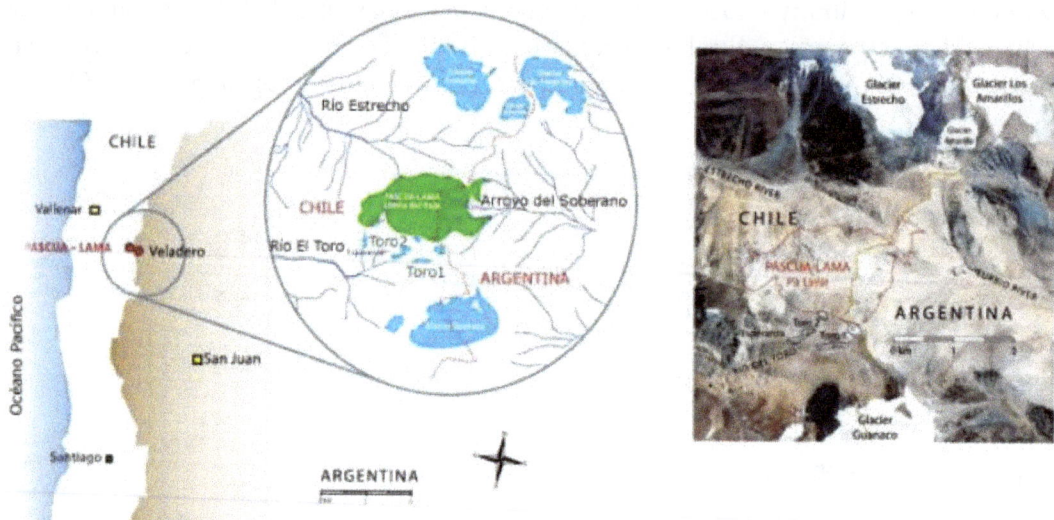

Figure 2. Glaciers and glacierets located in the study area

Natural Region	Region	Number of Glaciers	Surface [sqkm]
Far North	XV, I, II	28	42
Near North	III – IV	60	107
Central	V – VII, RM	1500	1019
South	IX, XIV, X	87	265
Far South	XI – XII	76	15460

Table 1. Geographical distribution of Glaciers in Chile

The current state of Chilean glaciers, according to [14] and [15], indicates that 87% is in decline, 6% in advance and 7% still remains unchanged.

Nowadays and in a near future, the water scarcity is a major concern all over the world. Especially in the arid to semi-arid mountainous regions, the local population hugely depends on alternative water resources such as those stored in glaciers or snow. Glaciological processes at high altitudes in such regions of Chile and Argentina (27°S to 33°S) have previously been studied with special focus on hydrology (e.g. [17]) and climatology (e.g. [17, 18]).

In the rivers that originate in the central Andes, the main water supply is generated from the snowfall that normally covers the upper mountain peaks each winter. Furthermore, besides these melting processes, surface runoff gradually flows into the rivers. During wet years, this runoff can deliver sufficient amount of water throughout the spring and summer period to compensate missing precipitation. In dry years, however, it tends to decrease towards the end of summer. In these drier periods, ice bodies have to provide enough melt water due to its natural resources accumulated in winter. The water contribution of glaciers, glacierets and other ice bodies is directly proportional to their area, as it is on the surface where melting caused by solar radiation, ambient heat and other environmental factors happens. As an example, the Chilean General Water Directorate (DGA – Dirección General de Aguas) estimates that the average, annual melting during summer fluctuates between 0.5 liters per second (l/s) per hectare and 1.8 l/s per hectare. The total area of the three glacierets located at the southern end of the study area was 16.5 hectares in March 2005, so that the total flow contributing to the basin was estimated to range from 8 l/s to 30 l/s during maximum of the melting period. In comparison, the average flow of Huasco River in Algodones, more westerly in an agricultural zone, is greater than 4000 l/s in summer.

In the Chilean Commune Alto del Carman, in the Atacama region, Figure 2, several glaciers can be found that have shown important variations during the past few decades. These glaciers namely are: Toro 1, Toro 2, Esperanza, Guanaco, Estrecho, Amarillo and Los Amarillos. Several other studies on glacier variations induced by climate have already been carried out in the past at different study sites, but in the vicinity of the mentioned glaciers. Ref. [19] studied the terminus of the Agua Negra Glacier (Argentine); [20] the Tronquitos glacier; [21] Cerro Topado; [22] the Huasco catchment. In 2003, [23] started a glacier and glaceriets (very small glaciers or ice masses of indefinite shapes in hollows and that have little or no movements; [24]) monitoring program in the region where all of the above mentioned glaciers are located.

1.1. Remote sensing of glaciers

Nowadays, traditional ground-based glacier monitoring studies can be complemented or even replaced by satellite-based data. Reflected solar radiation by the earth's surface is detected by optical, passive sensors in the visible (400–700 nm), near and short-wave infrared (700 nm – 7 μm) bands of the electromagnetic spectrum. Radiation emitted by the surface is detected by sensors in the thermal infrared (7 μm – 1 mm) bands. Electromagnetic radiation in the microwave bands (1 mm – 1 m) in remote sensing is used mostly by so-called synthetic aperture radar (SAR) active and passive systems [25].

Glacier monitoring can focus on several parameters such as glacier area and length, surface elevation, surface flow fields, accumulation and ablation rates or albedo. For mass balance study, in particular equilibrium line altitude (ELA), accumulation area ratio (AAR) and the mass balance gradient δb/δz are of importance [26].

All these parameters are relevant to study the influence of glaciers on the environment [27]. Important fields are:

- Glacier Geology: Bedrock material removed by glaciers is redistributed in the landscape. Erosion and deposition caused by glaciers forms U-shaped valleys, cirques, moraines and other glacial landforms. Glacial sediment is redistributed by wind or water, forming new soils and affecting the water quality of rivers, lakes and oceans.

- Glacier Hydrology: Glaciers store water during cool, wet winter periods and release it during warm, dry summer.

- Glacier Ecology: Glaciers are habitats for flora and fauna. Their meltwater provides aquatic habitat for endangered species.

- Glacier Hazards: Significant hazards are outburst floods, lahars, ice avalanches and spontaneous landslides.

Considering these two principal aspects, sensor types and glacier characteristics, one has to decide what kind of data has to be generated and which method is the most appropriate [28]. For example, digital elevation models are a key element for glacier volume change studies. Several conclusions can be drawn out of it, such as mass balance variability due to climate change. Furthermore, glacier surface flow velocity can be derived from differential InSAR observations and/or feature tracking in optical satellite images.

In scientific literature, a lot of examples can be reviewed that highlight the potential of remote sensing for glaciologic studies carried out all over the world, such as those in Alaska [29], Patagonia [30], the Andes [31], the Alps [32], the Himalaya [33] and Central Asia [34].

Table 2 gives an overview of most common satellites and their application in glacier monitoring.

Satellite	Operation dates	Bands	Glaciological application
Landsat	Series of satellites since 1972	VIS, IR	Glacier variations, spectral characteristics of snow and ice
SPOT	Series of satellites since 1986	VIS, IR	Glacier variations, spectral characteristics of snow and ice
SENTINEL	Series of satellites since 2014	VIS, IR, MW	Glacier variations, spectral characteristics of snow and ice, elevation change monitoring
CRYOSAT-2	Single mission since 2010	MW	Glacier variations, surface velocity estimation, elevation change monitoring
TERRA	Single mission since 2000	VIS, IR, MW	Glacier variations, elevation change monitoring

Table 2. Satellites and their possible application in glacier monitoring

2. Climate

Inter-annual global climate variability is mostly caused by the coupled ocean–atmosphere phenomenon ENSO (El Niño Southern Oscillation). It is a naturally, irregularly occurring phenomenon that is characterised by fluctuating ocean temperatures in the equatorial Pacific, between Australia and the west coast of South America. El Niño refers to a warming phase whereas La Niña refers to a cooling phase of the Pacific sea surface temperature (SST), the upper layer (0–10 cm) of the ocean. As ENSO is a coupled ocean–atmosphere phenomenon, air surface pressure in the tropical Western Pacific is higher than normal in case of El Niño and lower in case of La Niña. Both in general last several months or even years and vary in intensity.

ENSO has particular impact on inter-annual climate variability in Latin and South America. In Mexico and parts of the Caribbean, El Niño causes an augmentation in winter precipitation and a diminution in summer precipitation [35]. Severe droughts in Mexico have occurred during summer when El Niño was present [36]. La Niña, however, has an almost opposite effect, precipitation increases during summer months and decreases in winter.

In case of Colombia, El Niño causes reductions in precipitation, whereas La Niña is associated with stronger precipitation, which might result in floods [37]. Furthermore, they indicate that there exists a very high positive correlation between the Southern Oscillation Index (SOI) and river discharge in Colombia. During the December–January period, this relationship is stronger and weaker during April and May. There is also a regional difference. In western Colombia the influence of El Niño is stronger than in the east.

Large positive precipitation anomalies over the eastern part of the Andes (Ecuador and northern Peru) typically are observed during the warm episode [38].

In the Amazon region of Brazil northward to the Caribbean, deficiency in precipitation has been observed during El Niño [39]. In contrast, El Niño effects in southern Brazil are opposite to that in northeast Brazil and northern Amazonia. Positive and extremely large anomalies of rainfall have been observed during El Niño years [40, 41].

Between 30°S and 40°S, northern and central Chile and at high altitudes of the Andes in Argentina, most precipitation is recorded during winter months, with positive anomalies, which can be registered during early stages of El Niño. Due to the area's semi-arid conditions, local and regional economy might strongly be affected [42, 43, 44]. These events stand in contrast to what happens at low altitudes of Chile. Between 1991 and 1993, El Niño years, heavy rainfall triggered debris flows in Santiago de Chile, Antofagasta and surrounding areas [45].

At high altitudes of the Andes, large amounts of snow are consistently recorded. During summer, melting of accumulated snow is the main cause of river runoff. In Chile and central-western Argentina, north of 40°S, during El Niño years, streamflows were normal or above normal [46, 47]. In contrast, during La Niña years, negative anomalies of rainfall and snowfall can be observed with opposite consequences, which include below-normal summer streamflow. For this region, it is more probable that dry conditions occur during La Niña than wet conditions happen during El Niño years [42].

SST observations and analysis are often used to identify this oscillation and to predict the upcoming climate variability. Nevertheless, it has to be mentioned that it is the sub-surface ocean temperature, which indicates first an upcoming change, a transition from a cold to a warm phase or vice versa. It is important to understand that changes in sub-surface ocean temperatures are the first to respond to an oncoming change in the ENSO phase.

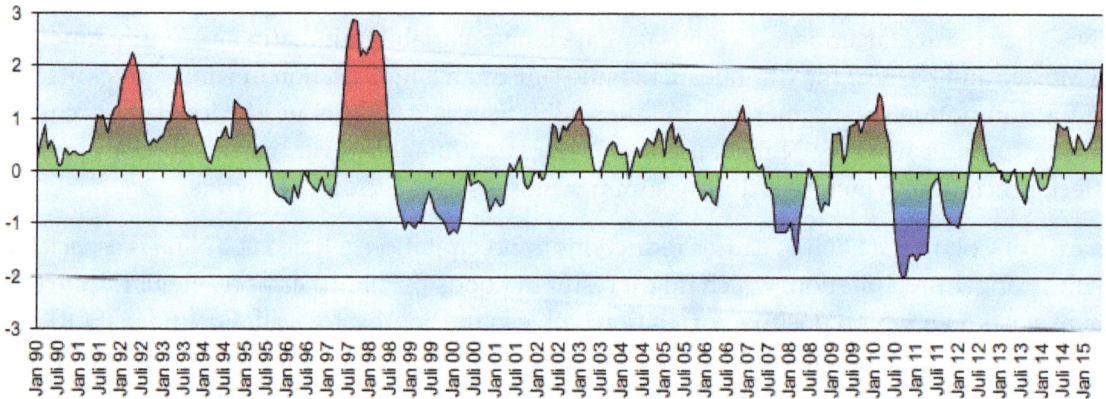

Figure 3. Multivariate ENSO Index (MEI) since 1990 (Source: [48])

Nevertheless, to monitor ENSO in literature the Multivariate ENSO Index (MEI), Figure 3, which is based on the six main observed variables over the tropical Pacific, can be found. These six variables are: sea-level pressure, zonal and meridional components of the surface wind, sea surface temperature, surface air temperature and total cloudiness fraction of the sky. At first, the MEI is computed separately for two consecutive months (Dec/Jan, Jan/Feb,...). Then, spatially filtering of the individual fields into clusters is applied [49]. The MEI is calculated afterwards as the first unrotated principal component (PC) of all six observed fields combined. Finally, the first PC on the co-variance matrix of the combined fields is extracted [50]. In Figure 4, the results of the past 25 years are illustrated. Positive values (in red) indicate El Niño phases whereas negative values (in blue) represent La Niña phases.

2.1. Remote sensing of climate

Since 1959 when the first space-borne observations of solar irradiance and cloud reflectance were made by the Vanguard-2 satellite, remote sensing gradually became a key observation and research method in climate change studies [51]. Satellite data of land, ocean and atmosphere are used to model and simulate the dynamics climate system in the past, present and future [52, 53].

Although satellite remote sensing, on a climate history time scale, is a relatively new technique, and therefore has some limitations such as short data spans of satellite records, biases associated with instruments and uncertainties in retrieval algorithms, it has to be considered in a series of particular applications that are listed in Table 3.

Application	Observation	Satellites
Global warming	Global temperature trends, particularly at the ocean surface and in the atmosphere.	NOAA, AQUA, TERRA
Snow and ice	Monitoring the dynamics of snow extent and ice covers	NOAA, SSM/I, ERS
Sea level change	Mapping of ocean surface topography	TOPEX/Poseidon, GRACE, Sentinel
Solar radiation	Determination of changes in the sun's luminosity	SORCE, Meteosat, Eumetsat
Aerosols	Atmospheric particles concentration	TERRA, AQUA
Water vapour and precipitaion	Precipitable water in the troposphere; spatial and temporal variability of precipitation at global scale	TERRA, AQUA
Clouds	Estimation of net cloud forcing	Cloudsat, TERRA, AQUA

Table 3. Remote sensing in climate change studies

3. Objectives and study area

The main objective of the present work was to determine surface area changes of glaciers in the Commune of Alto del Carmen (Chile) in the near north during the past two decades. And the specific goal of the study was to link these changes to climate variability in the study area.

Our study focuses on the following glaciers and glacierets located in the Commune of Alto del Carmen: Toro 1, Toro 2, Esperanza (all defined as glacierets) Guanaco, Estrecho, Amarillo and Los Amarillos (all defined as glaciers), see Figure 2. Ref. [22] found out that their spatial distribution is highly correlated with natural factors, such as terrain characteristics, solar radiation and shadowing effects. Ref. [54] indicates that only little ice flow exists and that surface areas are smaller than 2 sqkm in 2007. Furthermore, they mention, based on ground penetration radar (GPR) measurements, that ice thickness can reach up to 100 meters. More historical studies of area changes [23] have shown that during the past 50 years, surface area has reduced significantly in almost all glaciers and glacierets.

Furthermore, it is noteworthy that an important mining project of gold, silver, copper and minerals is developed in the nearby vicinity. At first, the responsible mining company (Barrick Gold) proposed to move Toro 1, Toro 2 and Esperanza glacier, as they were considered of little relevance to the basin water, are located in the pit area of the foreseen mine and were in process of disappearance (observed high melt rates). National and Regional Environmental commissions (CONAMA and COREMA, respectively) approved the mining project in general. Nevertheless, they made a couple of observations to the initial proposal and prohibited to move glaciers. As a consequence to the already carried out exploitation activities, such as drilling, and installation of infrastructure, e.g. access roads, several negative effects can already be observed.

• Watershed: There are two rivers, which are directly affected by the Pascua Lama mining project, as their source is in that area and are fed by melting snow that infiltrates the ground: On the Argentinian side, it is river Turbio and on the Chilean side, it is El Estrecho river. On its way through rocks and materials that make up the upper river sources, the water comes into contact with the minerals in the soils of the Pascua-Lama deposit. Due to its chemical and mineralogical composition, water becomes acid and can dissolve metals contained in the rocks.

• Glacier: In 2001, COREMA approved the so-called Environmental Impact Assessment of the Pascua Lama Project, which was presented by Barrick Gold. In this study report, several environmental impacts were outlined. Furthermore, several mitigation measures and monitoring programs were proposed to protect the environment. The assessment covered those environmental components, which were considered as most relevant, including: air quality; levels of noise and vibration; flow and quality of surface and groundwater; geomorphology, drainage and soil; vegetation and flora; terrestrial fauna; aquatic flora and fauna; landscape; cultural heritage; road service levels; and socio economy. Nevertheless, glaciers were not considered to be protected. It was proposed to move them to a different location. As this was never realized, the mining company constructed a transportation road right through one of them.

4. Methodology and data

Glacier and glacieret surface area were determined from georeferenced satellite images acquired by Landsat 5, 7 and 8. These satellite images were acquired during summer months from 1994 until present, for two reasons: (1) Cloud cover in the satellite images had to be less than 10% and (2) glaciers can be detected very easily and with high certainty as seasonal snow cover does not exist anymore. Their ground sampling distance (GSD) is 30 m in all spectral bands. Landsat 7 satellite images with SLC (Scan Line Correction) off had to be re-processed to fill gaps. This was done by spectral interpolation for every affected satellite image.

Satellite	Time Span	Total Number of images
Landsat 5	1994-2003	10
Landsat 7	2004-2013	10
Landsat 8	2014-2015	2

Table 4. Satellite images used in this study

With all the satellite images acquired, corrected and georeferenced, at first, a visual identification, interpretation and analysis of the glaciers and glacierets were carried out to detect inter-annual variations. In a second processing step, image digitalization was applied with the aim to measure glacier extent and to derive surface area variations.

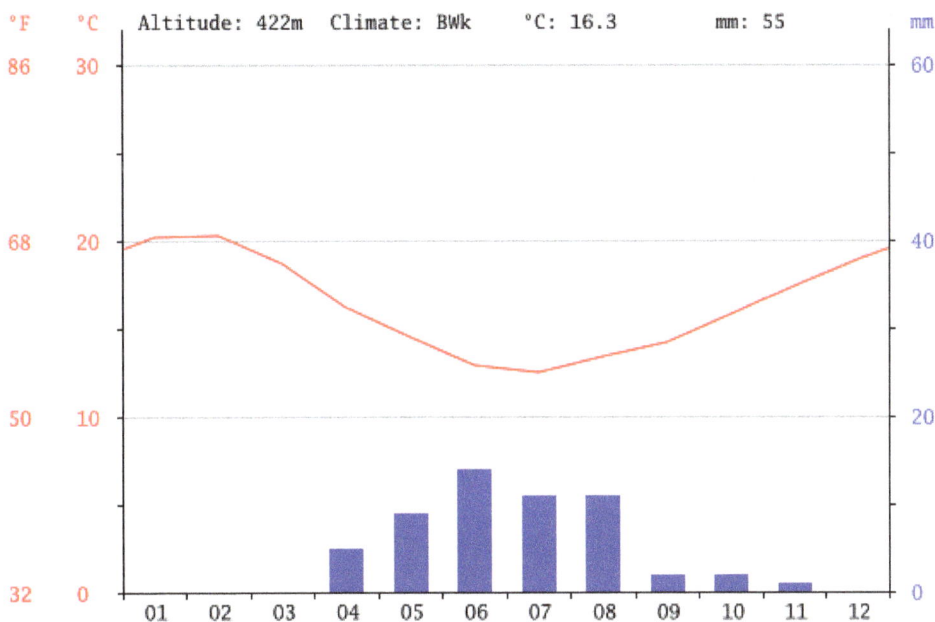

Figure 4. Climate chart for the city of Vallenar

Climate data, which is relevant for this study, is available for free on the internet. There are several weather stations situated in the Atacama region close to the study area. The most complete data set of precipitation and temperature records is available for the city of Vallenar, located at 100 km to the northwest of the study area. Furthermore, climate data registered for the cities of Conay (60 km to the north) and Chollay (50 km to the northwest) are available, but only from 2008 onwards and with some gaps. Therefore, measurements until 2008 at a fourth station, namely La Olla (80km to the southwest), are taken into account. This data was taken from [23].

In a final step, Pearson product–moment correlation between climate and digitised data is generated. Both are linked together with a special focus on anomalies and unusual events, e.g. high temperature events, heavy and prolonged rainfall. The statistical correlation is determined by the relationship or dependence between the two studied variables: area (sqkm), temperature (T °), precipitation (mm), in a two-dimensional distribution. In case that there can be found one of these variables influencing another, it can be stated that the variables are correlated or that there exists correlation between them.

The linear correlation coefficient is calculated as follows:

$$r = r_{xy} / r_x r_y \; ; r \; \varepsilon \left[-1,1 \right]$$

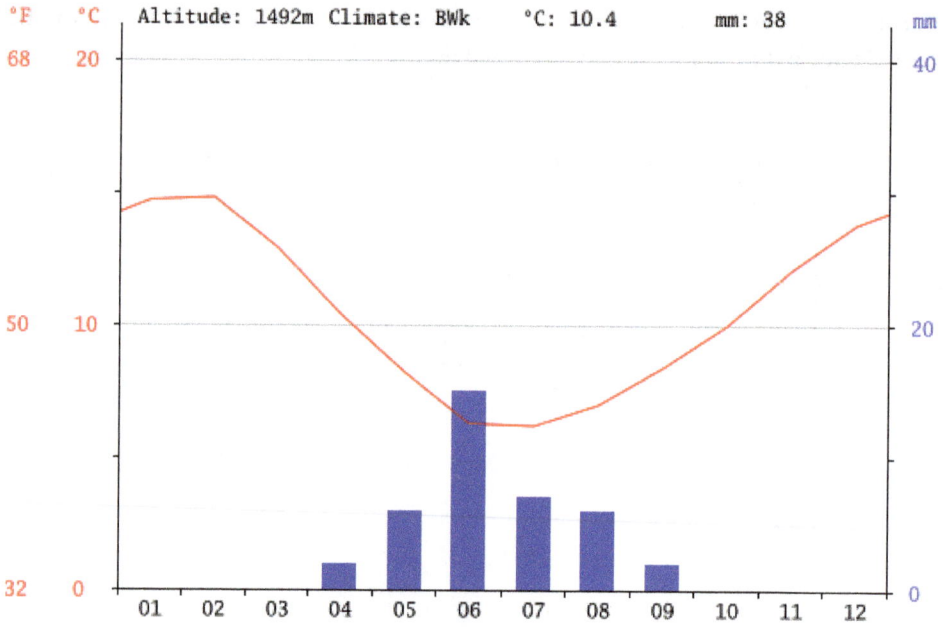

Figure 5. Climate chart for the city of Chollay

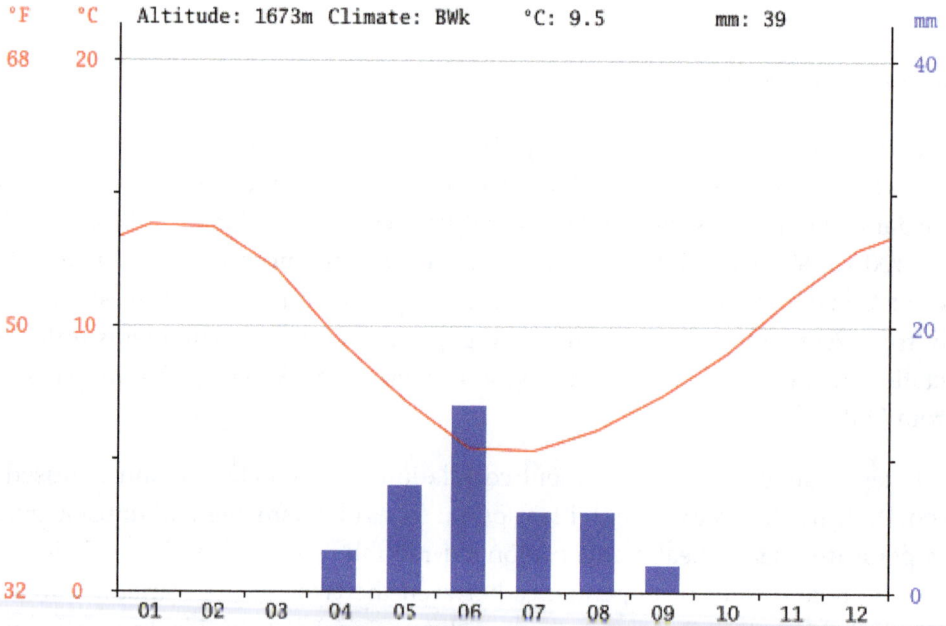

Figure 6. Climate chart for the city of Conay

where r_{xy} is the xy covariance, r_x and r_y are the standard deviations, respectively.

ϱ can vary between −1 and 1. In case that the correlation coefficient is 1, there is a perfect increasing linear correlation; if it is −1, a perfect decreasing linear relationship can be found between the variables; in all other cases, the value indicates to which degree a linear dependency exists. So, a 0 value means that no correlation exists between two analysed parameters.

For our study, this can be interpreted the following way:

	$r_{surface, temperature}$	$r_{surface, precipitation}$
Positive	Glacier surface area decrease (increase) due to temperature decrease (increase)	Glacier surface area decrease (increase) due to precipitation decrease (increase)
Zero	No correlation between glacier surface area and temperature variations	No correlation between glacier surface area and temperature variations
Negative	Glacier surface área decrease (increase) due to temperature increase (decrease)	Glacier surface área decrease (increase) due to precipitation increase (decrease)

Table 5. Possible results of correlation coefficients statistics and their interpretation regarding surface

5. Results and discussion

The Figures 7–13 show how surface area of the studied glaciers and glacierets has reduced during the past 25 years. The blue line indicates the glacier extent in 1994, whereas the green line delimitates the glacier area which was observed in the last suitable satellite image. The surface area which was lost during the past two decades is highlighted by linear hatching in red colour.

In case of Los Amarillos glacier (Figure 7), surface area difference between 1994 and 2015 is – 43.53%. More area loss can be identified in the southeastern part of the glacier.

Amarillo glacier (Figure 8), during the past 20 years, has diminished 62.12%. In particular, the southern extend of the glacier has significantly reduced.

Figure 9, which shows the surface area loss at Estrecho glacier, indicates that it has lost surface area almost uniformly; no specific spatial trend can be identified. All margins have uniformly reduced by means of extend. Surface area loss was 43.45%.

For Guanaco glacier (Figure 10), a similar behaviour as for Estrecho glacier can be identified, except the most westerly area, which has completely gone. In 1994, it was already isolated and only connected to the main glacier area by an ice bridge. Total surface area loss during the past 2 decades was 33.84%.

Toro 1 and Toro 2 (Figures 11 and 12) have shown similar spatial behaviours. The eastern part of both glacierets is gone. Only about 4% of the original surface area remains.

In Esperanza glacier (Figure 13), whose spatial extend was from north to south, surface area loss happened mostly in this spatial direction. Nowadays, the geometry of this glacieret is almost circular.

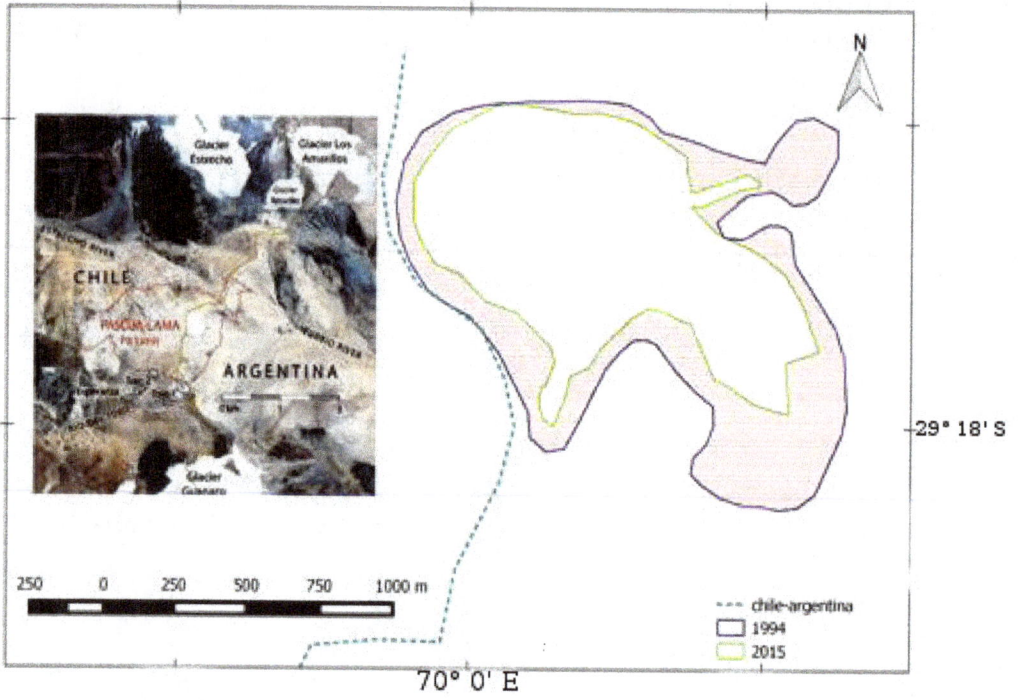

Figure 7. Surface area change of glacier Los Amarillos

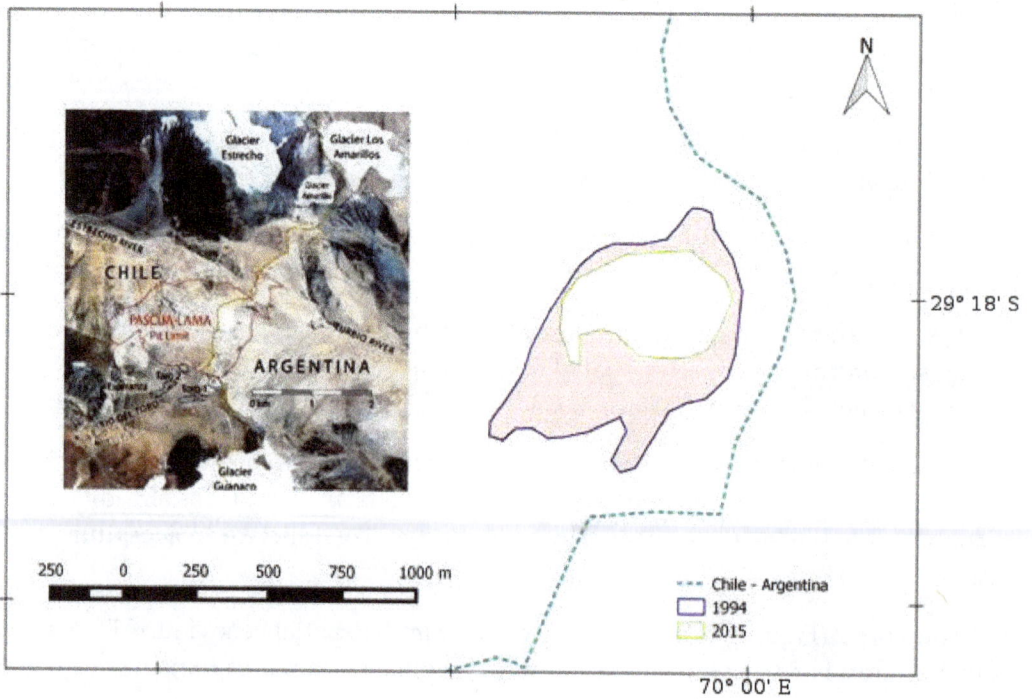

Figure 8. Surface area change of glacier Amarillo

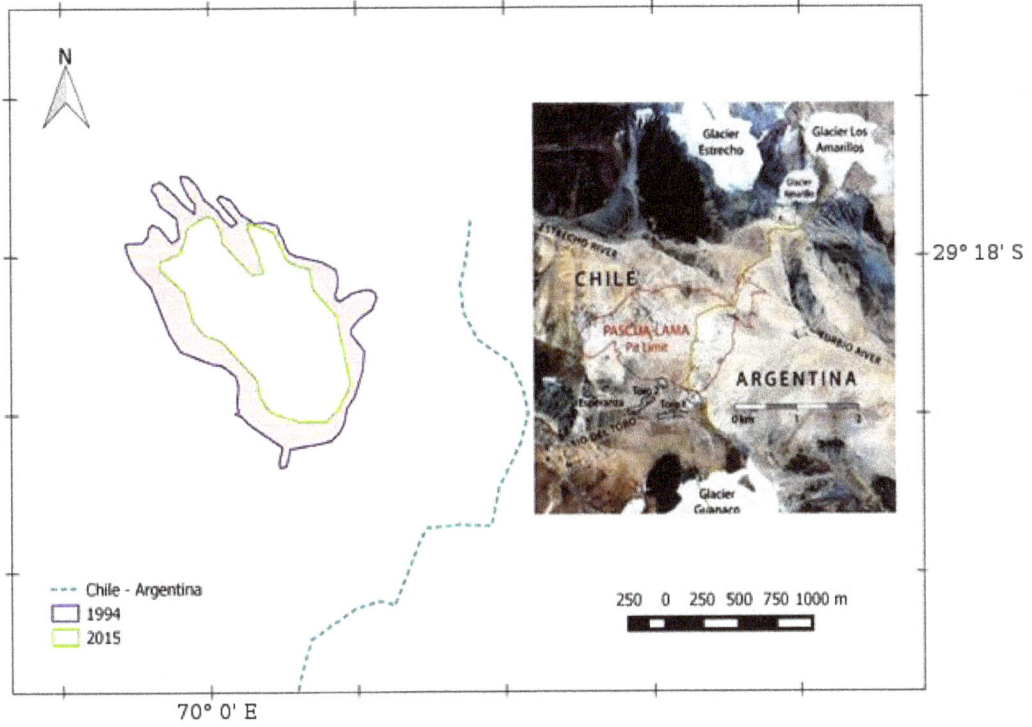

Figure 9. Surface area change of glacier Estrecho

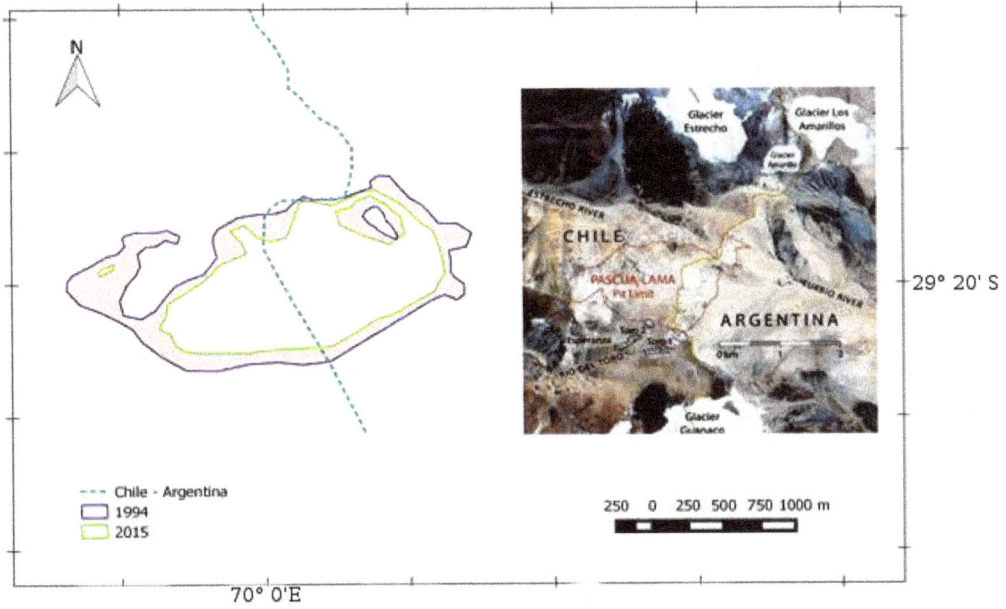

Figure 10. Surface area change of glacier Guanaco

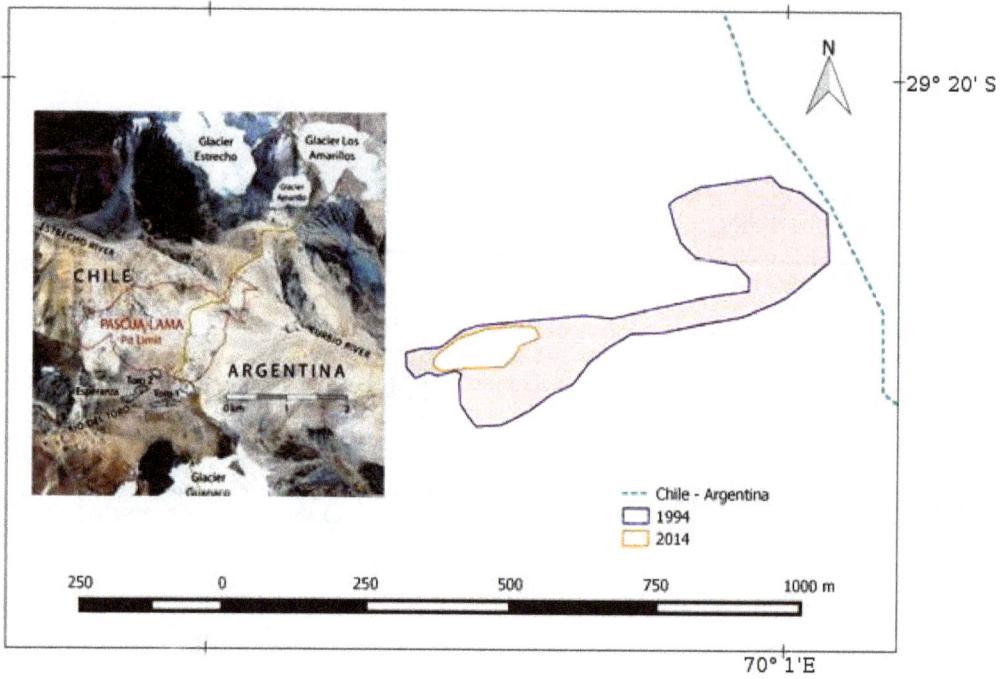

Figure 11. Surface area change of glacieret Toro 1

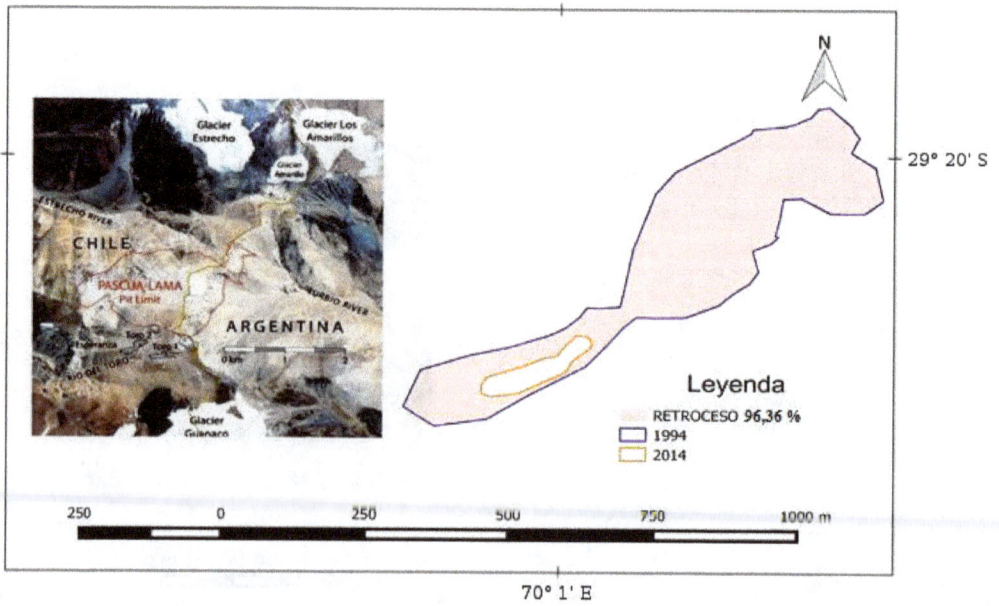

Figure 12. Surface area change of glacieret Toro 2

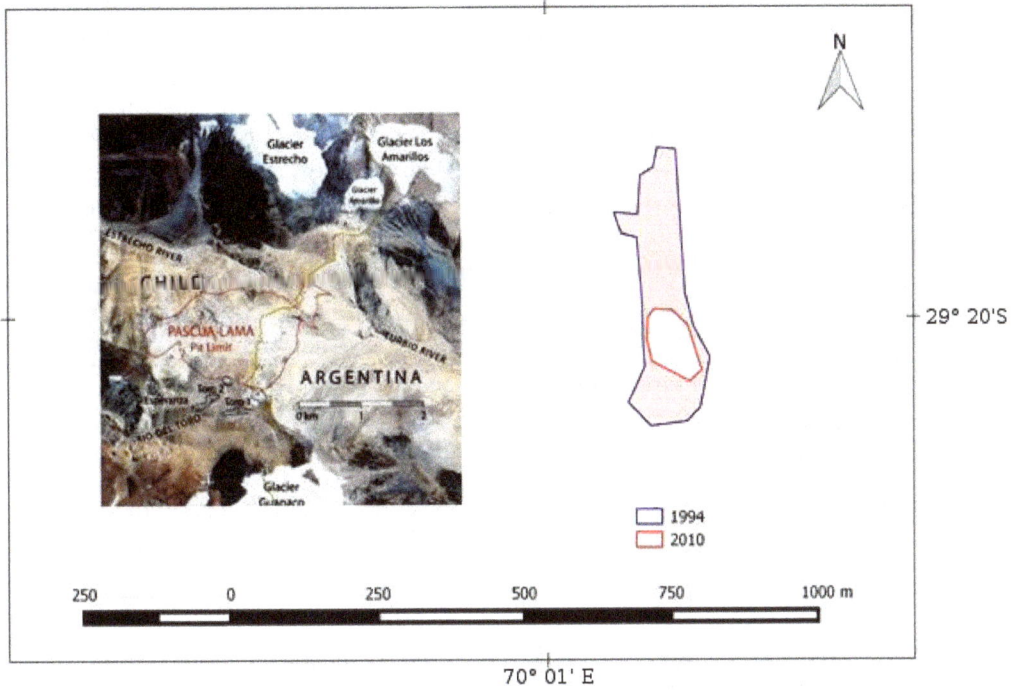

Figure 13. Surface area change of glacieret Esperanza

Table 6 shows the results of the digitalisation carried out to determine surface area variations. Only summer months were considered and satellite images of every second year were taken into account. The generated graphs show that all glaciers and glacierets have suffered surface area loss during the past 21 years. Nevertheless, some of them were able to recover during 2004, 2006 and/or 2008.

	1994	1996	1998	2000	2002	2004	2006	2008	2010	2012	2014	2015
Los Amarillos	1,291	1,104	1,293	1,296	1,149	1,263	1,031	1,274	0,855	0,796	0,852	0,729
Amarillo	0,378	0,376	0,491	0,379	0,331	0,317	0,334	0,278	0,168	0,134	0,173	0,143
Estrecho	1,459	1,865	1,366	1,311	1,081	1,356	1,439	1,350	1,117	0,954	0,986	0,825
Guanaco	2,077	2,254	2,181	2,038	1,999	1,942	1,720	1,893	1,668	1,479	1,482	1,374
Toro 1	0,163	0,137	0,233	0,101	0,059	0,063	0,090	0,084	0,019	0,000	0,008	0,000
Toro 2	0,199	0,124	0,364	0,044	0,022	0,060	0,089	0,113	0,000	0,000	0,007	0,000
Esperanza	0,036	0,015	0,106	0,009	0,025	0,018	0,079	0,043	0,009	0,000	0,056	0,034

Table 6. Absolute surface area values in sqkm from 1994 till 2015

Toro 1, Toro 2 and Esperanza have not been gone. Here, zero values indicate that it was not possible to detect them in the satellite images of the corresponding years due to spatial resolution.

Since 1994, surface area of glaciers and glacierets analysed in this study has reduced as shown in Table 7. Toro 1 and Toro 2 have lost almost its entire surface. Esperanza seems to be stable as it was able to recover during the past decade and although it has lost 5% of its surface between 1994 and 2015. All the other glaciers show a continuous trend of surface area loss during the past 20 years.

Glacier	Glacier surface area sqkm			Loss between 1994 and last year
	1994	2004	Last year	
Los Amarillos	1,291	1,031	0,729	43,53%
Amarillo	0,378	0,317	0,143	62,12%
Estrecho	1,459	1,356	0,825	43,45%
Guanaco	2,077	1,942	1,374	33,84%
Toro 1	0,163	0,063	0,008	95,10% (until 2014)
Toro 2	0,199	0,060	0,007	96,48% (until 2014)
Esperanza	0,036	0,018	0,034	5,60%

Table 7. Surface aera loss over the last 2 decades

In order to characterise the temporal behaviour of the glacier surface area lost and to detect its correlations with rainfall and temperature, trends and correlation coefficients as shown in Table 8 were calculated.

	Trend (sqkm/a)	$r_{Surface, Temperature}$	$r_{Surface, Precipitation}$
Los Amarillos	-0,024	-0,4	0,2
Amarillo	-0,015	-0,4	0,4
Estrecho	-0,032	-0,4	0,8
Guanaco	-0,039	-0,5	0,5
Toro 1	-0,009	-0,2	0,5
Toro 2	-0,010	0,0	0,4
Esperanza	-0,001	0,4	0,1

Table 8. Trend in surface area lost and correlation coefficients

Precipitation at the mentioned climate stations Conay, Chollay and Vallenar are shown in Table 9. A couple of peaks can be observed which in general coincide with El Niño years.

Furthermore, compared to long-term observations as shown in Figures 6, 7 and 8, in 2004, 2006, 2012 and 2015, an absence of precipitation is observable. This also coincides with La Niña events that have been reported. During the past two decades, precipitation only in 1994 (Vallenar) and 2014 (Vallenar, Conay, Chollay) reached normal values.

Station	1994	1996	1998	2000	2002	2004	2006	2008	2010	2012	2014	2015
Vallenar	24,885	5,27	0,125	1,965	0,763	0,000	0,000	7,900	4.875	0,100	12.300	0,575
Conay	--	--	--	--	--	--	--	8,150	5,400	0,050	12,100	0,000
Chollay	--	--	--	--	--	--	--	8,15	5,525	0,175	11,65	0,525

Table 9. Average precipitation in mm during winter months

At all three stations, a positive temperature trend can be observed. Table 10 shows mean temperature during summer from 1994 onwards at Vallenar and from 2010 onwards for Conay and Chollay. In particular, summer of 2013 and 2015 shows huge variations in comparison to long-term climate observation (Figures 6, 7 and 8). Summer (December, January, February) mean temperature observed between 1982 and 2012 (Source: http://en.climate-data.org/) Vallenar station is 19.8°C; at Conay station 13.5°C; at Chollay station 14.4°C.

Our results indicate that all studied ice bodies have reduced in terms of extent over the past few decades and that there is a significant surface area loss in the glaciers and glacierets studied. Some of them show significant changes (Guanaco, Estrecho and Los Amarillos), whereas others seem to be more stable (Esperanza). Smaller ice bodies, glacierets, are more affected than glaciers. In particular, the Guanaco glacier shows major loss of surface area with a trend of –0.039 sqkm/a. It has already been reported by [55] that Guanaco, in comparison to other glaciers and glacierets in this area, shows major melt rates during summer months. On the other hand, the Esperanza glacier is the one which has almost remained stable. Its surface area loss has a tendency of –0.001 sqkm/a.

Station	1994	1996	1998	2000	2002	2004	2006	2008	2010	2011	2012	2013	2014	2015
Vallenar	19,7	19,9	21,0	19,4	19,3	17,2	--	--	18,3	19,0	21,5	22,7	22,3	23,3
Conay	--	--	--	--	--	--	--	--	19,6	18,5	20,7	22,6	20,1	24,4
Chollay	--	--	--	--	--	--	--	--	19,5	19,6	20.5	22,7	20,6	24,3

Table 10. Average temperatura in °C during summer months

In addition, considering climate variables like temperature and precipitation, surface area variation of glaciers and glacierets does weakly correlate with them. Only in case of Guanaco glacier, where the coefficient of correlation between surface temperature and surface-precipitation reaches values around 0.5, relationship between the variables can be interpreted as more dominant.

ENSO events can clearly be identified using MEI. Although in case of the Chilean Andes region, it is supposed that during El Niño years precipitation increases (as in 1998), this does not mean that glaciers and glacierets do not suffer surface area loss. But this does not mean that during La Niña years (e.g. 2010/11), surface area loss does accelerate neither.

A possible negative impact on the glaciers caused by the Pascua-Lama mining project or any other human activity was not considered, and our results do not indicate any correlation either. This coincides with a scientific report, which was recently published by [56]. They also could not find any evidence for surface area loss due to anthropogenic intervention in the study area. They attribute glacier variability to climate change and ENSO events. Nevertheless, due to contentious issues, this subject has to be analysed apart.

6. Conclusions

It has to be concluded that the overall climate situation at high latitudes of the Chilean Andes Mountain does have a negative impact on ice bodies. La Niña and El Niño events can be detected and it is possible to correlate them with variations in temperature and precipitation. As temperature during the past 25 years has augmented and precipitation has decreased, glaciers and glacierets have diminished their surface area.

There is evidence that lack of precipitation has a major impact on surface area loss. As an example, Esperanza glacieret was able to recover surface area loss in 2014 when precipitation average at all three stations was almost normal. This coincides with the results reported by [23]. Nevertheless, our results also indicate huge temperature variations and therefore we conclude that nowadays both climate variables have to be considered as responsible for glacier and glacieret surface area loss in the Commune of Alto del Carmen.

Although anthropogenic interventions in the study area are present, such as Pascua-Lama mining project, climate variability does play a major role in glacier changes. Nevertheless, this does not mean that human interference on ecosystems has to be tolerated. Every man-made alteration has to be reviewed critically and in particular in case of exploitation of natural resources, vital for flora and fauna, they have to be carried out complying with several standards protecting our planet. Nowadays, there are lots of technical possibilities, such as GIS and remote sensing that are well known and understood from geosciences, which allow a sustainable management of natural resources.

Author details

Guido Staub* and Catherinne Muñoz

*Address all correspondence to: gstaub@udec.cl

Universidad de Concepción, Departamento de Ciencias geodésicas y Geomática, Los Ángeles, Chile

References

[1] Marshall, S. Glacier retreat crosses a line. Science. 2014;872:345.

[2] Mote, P. and Kaser, G. The Shrinking Glaciers of Kilimanjaro: Can Global Warming Be Blamed?. American Scientist. 2007;95:318.

[3] IPCC. CLIMATE CHANGE 2001: THE SCIENTIFIC BASIS [Internet]. 2001. Available from: http://www.ipcc.ch/ipccreports/tar/wg1/ [Accessed: 01.09.2015]

[4] The Core Writing Team, Pacauri, R. & Reisinger, A., editors. Climate Change 2007: Synthesis Report - Contribution of Working Groups I, II and III to the Fourth Assessment Report of the Intergovernmental Panel on Climate Change. Geneva, Switzerland: IPCC; 2007. 104 p.

[5] Painter, T., Flanner, M., Kaser, G., Marzeion, B., VanCuren, R. and Abdalati, W. End of the Little Ice Age in the Alps forced by industrial black carbon. Proceedings of the National Academy of Sciences of the United States of America. 2013;110(38):15216–15221.

[6] Shakun, J., Clark, P., He, F., Lifton, N., Li, Z., Otto-Bliesner, B. Regional and global forcing of glacier retreat during the last deglaciation. Nature Communications. 2015;6DOI: 10.1038/ncomms9059

[7] López-Morenoa,, J., Fontanedaa, S., Bazob, J., Revueltoa, J., Azorin-Molinaa, C., Valero-Garcésa, B., Morán-Tejedaa, E., Vicente-Serranoa, S., Zubietac, R., Alejo-Cochachínd, J. Recent glacier retreat and climate trends in Cordillera Huaytapallana, Peru. Global and Planetary Change. 2014;112:2014. DOI: 10.1016/j.gloplacha.2013.10.010

[8] van der Ploeg, F. Natural Resources: Curse or Blessing?. Journal of Economic Literature. 2011;49(2):366-420.

[9] O'Neill, D. Measuring progress in the degrowth transition to a steady state economy. Ecological Economics. 2012;84:221–231.

[10] Harris, J., Roach, B. Environmental and Natural Resource Economics: A Contemporary Approach. 3rd ed. Routledge; 2014. 584 p.

[11] American Chemical Society. What are the greenhouse gas changes since the Industrial Revolution? [Internet]. Available from: http://www.acs.org/content/acs/en/climatescience/greenhousegases/industrialrevolution.html [Accessed: 02.09.2015]

[12] Delworth, T. and Knutson, T. Simulation of Early 20th Century Global Warming. Science. 2000;287:2246–2250.

[13] Zhang, R., Delworth, T. Simulated Tropical Response to a Substantial Weakening of the Atlantic. Journal of Climate. 2015;18:1853–1860.

[14] NSIDC. World Glacier Inventory [Internet]. 1999 [Updated: 2015]. Available from: https://nsidc.org/data/docs/noaa/g01130_glacier_inventory/ [Accessed: 01.10.2015]

[15] Glaciares: Reservas Estratégicas de Agua Dulce [Internet]. 2011. Available from: http://es.slideshare.net/boletinvertientes/glaciares-reservas-estratgicas-de-agua-dulce-chile-sustentable [Accessed: 05.09.2015]

[16] Favier, V., Falvey, M., Rabatel, A., Praderio, E., and López, D. Interpreting discrepancies between discharge and precipitation in high-altitude area of Chile's Norte Chico region (26°–32°S). Water Resources. 2009;45DOI: 10.1029/2008WR006802

[17] Masiokas, M. H., Villalba, R., Luckman, B. H., Le Quesne, C., and Aravena, J. C. Snowpack variations in the Central Andes of Argentina and Chile, 1951–2005: large-scale atmospheric influences and implications for water resources in the region. Journal of Climate. 2006;19:6334-6352.

[18] Vuille, M. and Milana, J. P. High-latitude forcing of regional aridification along the subtropical west coast of South America. Geophysical Research Letters. 2007;34DOI: 10.1029/2007GL031899

[19] Leiva, J. C. Recent fluctuations of the Argentinian glaciers. Global and Planetary Change. 1999;22:169–177. DOI: 10.1016/S0921-8181(99)00034-X

[20] Rivera, A., Acuña, C., Casassa, G., and Bown, F. Use of remote sensing and field data to estimate the contribution of Chilean glaciers to sea level rise. Annals of Glaciology. 2002;34:367–372.

[21] Ginot, P., Kull, C., Schotterer, U., Schwikowski, M., and Gaeggeler, H. W. Glacier mass balance reconstruction by sublimation induced enrichment of chemical species on Cerro Tapado (Chilean Andes). Climate of the past. 2006;2:21–30.

[22] Nicholson. L., Maríin, J., Lopez, D., Rabatel, A., Bown, F., and Rivera, A. Glacier inventory of the upper Huasco valley, Norte Chico, Chile: glacier characteristics, glacier change and comparison to central Chile. Annals of Glaciology. 2009;50:111–118.

[23] Rabatel, A., Castebrunet, H., Favier, V., Nicholson, L., and Kinnard, C. Glacier changes in the Pascua Lama region, Chilean Andes (29°S): recent mass balance and 50yr surface area variations. The Cryosphere. 2011;5:1029–1041.

[24] Kumar, R. Encyclopaedia of snow, ice and glaciers, Encyclopaedia of earth sciences series. 2014. 436 p.

[25] Albertz, J. Einführung in die Fernerkundung. Grundlagen und Interpretation von Luft- und Satellitenbildern. 3rd ed. WBG; 2007. 264 p.

[26] Dyurgerov M. B., Meier M. F. Twentieth century climate change: Evidence from small glaciers. PNAS. 2000;97(4):1406–1411.

[27] Karpilo, R. Glacier monitoring techniques. In: Young, R., Norby, L, editors. Geological Monitoring: Boulder, Colorado. The Geological Society of America; 2009. p. 141-162.

[28] Sulzer, W., Lieb, G.. Die Gletscher im wandel der Zeit - Gletschermonitoring am Beispiel der Pasterze. Vermessung & Geoinformation. 2009;3:371–382.

[29] Arendt, A., Echelmeyer K.A., Harrison W.D., Lingle C.S., Valentine V.B. Rapid wastage of Alaska glaciers and their contribution to rising sea level. Science. 2002;297(5580):382–386.

[30] E. Rignot, A. Rivera, G. Casassa. Contribution of the Patagonia Icefields of South America to sea level rise. Science. 2003;302(5644):434–437.

[31] Racoviteanu A., Arnaud Y., Williams M. Decadal changes in glacier parameters in Cordillera Blanca, Peru derived from remote sensing. Journal of Glaciology. 2008;54(186):499–510.

[32] Paul F., Kääb A., Maisch M., Kellenberger T., Haeberli W. Rapid disintegration of Alpine glaciers observed with satellite data. Geophysical Research Letter. 2004;31(21):1–4.

[33] Kääb A. Combination of SRTM3 and repeat ASTER data for deriving alpine glacier flow velocities in the Bhutan Himalaya.. Remote Sensing of the Environment. 2005;94(4):463–474.

[34] Bolch T. Climate change and glacier retreat in northern Tien Shan (Kazakhstan/Kyrgyzstan) using remote sensing data. Global Planet Change. 2007;56(1–2):1–12.

[35] Magaña V. and Quintanar, A.. On the use of a general circulation model to study regional climate. In: Cambridge University Press, editor. 2nd. UNAM-CRAY Supercomputing Conference on Earth Sciences; Mexico City. 1997. pp. 39–48.

[36] Magaña, V., Pérez, L. and Conde, C. El fenómeno de El Niño y la Oscilación del sur y sus impactos en México. Revista Ciencias. 1998;51:14–18.

[37] Poveda, G. and Mesa, O. Feedbacks between hydrological processes in tropical South America and large scale oceanic-atmospheric phenomena. Journal of Climate. 1997;10:2690–2702.

[38] Bookhagen, B. and Strecker, M. Modern Andean rainfall variation during ENSO cycles and its impact on the Amazon drainage basin. In: C. Hoorn and F.P. Wesselingh, editors. Amazonia, Landscape and Species Evolution: A Look into the Past. 1st ed. 2010.

[39] Diaz, H. and Kiladis, G. Atmospheric teleconnections associated with the extreme phases of the Southern Oscillation. In: H.F. Diaz and V. Markgraf, editors. Paleoclimatic Aspects of El Niño/Southern Oscillation. Cambridge Press; 1997.

[40] Grimm, A., Barros, V. and Doyle, M. Climate Variability in Southern South America Associated with El Niño and La Niña Events. Journal of Climate. 2000;13:35–58.

[41] Halpert, M. and Ropelewski, C. Surface Temperature Patterns associated with the Southern Oscillation. Journal of Climate. 1992;5:577–593.

[42] Compagnucci, R. H. and Vargas, W. M. Inter-annual variability of the Cuyo rivers' streamflow in the Argentinean Andean mountains and ENSO events. International Journal of Climatology. 1998;18:1593–1609.

[43] Rutland J. and Fuenzalida, H. Synoptic aspects of the central Chile rainfall variability associated with the Southern Oscillation. International Journal of Climatology. 1991;11:63–76.

[44] Quinn, W. And Neal, V. The historical record of El Niño events. Climate since A.D. 1500. In: S. Bradley and P. D. Jones, editors. Routledge; 1982.

[45] Garreaud, R. and Rutllant, J. Análisis meteorológico de los aluviones de Antofagasta y Santiago de Chile en el periodo 1991-1993. Atmósfera. 1996;9:251–271.

[46] Waylen, P. and Caviedes, C. Annual and seasonal fluctuations of precipitation and streamflow in the Aconcagua river basin, Chile. Journal of Hydrology. 1992;120:79–102.

[47] Martinez, C., Fernández, A. and Rubio, P. Flow and climatic variability on a South American Mid-Latitude basin: Río Aconcagua, Central Chile (33°S). Boletín de la Asociación de Geógrafos Española. 2012;58:481–485.

[48] Baldenhofer, K. ENSO Lexikon [Internet]. 1998 [Updated: 2015]. Available from: http://www.enso.info/enso-lexikon/lexikon-m.html [Accessed: 02.09.2015]

[49] Wolter, K. The Southern Oscillation in surface circulation and climate over the tropical Atlantic, Eastern Pacific, and Indian Oceans as captured by cluster analysis. Journal of Applied Meteorology and Climatology. 1987;26:540–558.

[50] Wolter, K., and Timlin, M. Monitoring ENSO in COADS with a seasonally adjusted principal component index. In: Proc. of the 17th Climate Diagnostics Workshop; Norman, OK, USA. 1993. pp. 52–57.

[51] Li, J., Wang, M. H., Ho, Y. S. Trends in research on global climate change: A Science Citation Index Expanded-based analysis. Global Planet Change. 2011;77:13–20.

[52] Ghent, D., Kaduk, J., Remedios, J., Balzter, H. Data assimilation into land surface models: The implications for climate feedbacks. International Journal of Remote Sensing. 2011;32:617–632.

[53] Saha, S., Moorthi, S., Pan, H.-L., Wu, X., Wang, J., Nadiga, S., et al. The NCEP Climate Forecast System Reanalysis. Bull. Amer. Meteor.. 2010;91(8):1015–1057.

[54] GolderAssociates S.A. Área de Pascua Lama,Tercera Región de Atacama, Recopila-ción de estudios de línea base actualizada de la criósfera. In: 2009.

[55] MacDonell, S., Mölg, T., Nicholson, L. Kinnard, C. The surface energy balance of the Guanaco and Toro 1 glaciers in the Norte Chico region, Chile. In: EGU General As-sembly 2010; 2-7 May, 2010; Vienna, Austria. COPERNICUS; 2012. p. 11042.

[56] Bown, F., Carrión, D., Bravo, C., Hernández, J., Muñoz, C., Correa, J., et al. Green-peace [Internet]. 2015. Available from: http://www.greenpeace.org/chile/Global/chile/Fotos/Clima%20y%20Energia/2015/10/Informe%20Excedencia_2015-2.pdf [Accessed: November 2015]

Permissions

All chapters in this book were first published in EARS, by InTech Open; hereby published with permission under the Creative Commons Attribution License or equivalent. Every chapter published in this book has been scrutinized by our experts. Their significance has been extensively debated. The topics covered herein carry significant findings which will fuel the growth of the discipline. They may even be implemented as practical applications or may be referred to as a beginning point for another development.

The contributors of this book come from diverse backgrounds, making this book a truly international effort. This book will bring forth new frontiers with its revolutionizing research information and detailed analysis of the nascent developments around the world.

We would like to thank all the contributing authors for lending their expertise to make the book truly unique. They have played a crucial role in the development of this book. Without their invaluable contributions this book wouldn't have been possible. They have made vital efforts to compile up to date information on the varied aspects of this subject to make this book a valuable addition to the collection of many professionals and students.

This book was conceptualized with the vision of imparting up-to-date information and advanced data in this field. To ensure the same, a matchless editorial board was set up. Every individual on the board went through rigorous rounds of assessment to prove their worth. After which they invested a large part of their time researching and compiling the most relevant data for our readers.

The editorial board has been involved in producing this book since its inception. They have spent rigorous hours researching and exploring the diverse topics which have resulted in the successful publishing of this book. They have passed on their knowledge of decades through this book. To expedite this challenging task, the publisher supported the team at every step. A small team of assistant editors was also appointed to further simplify the editing procedure and attain best results for the readers.

Apart from the editorial board, the designing team has also invested a significant amount of their time in understanding the subject and creating the most relevant covers. They scrutinized every image to scout for the most suitable representation of the subject and create an appropriate cover for the book.

The publishing team has been an ardent support to the editorial, designing and production team. Their endless efforts to recruit the best for this project, has resulted in the accomplishment of this book. They are a veteran in the field of academics and their pool of knowledge is as vast as their experience in printing. Their expertise and guidance has proved useful at every step. Their uncompromising quality standards have made this book an exceptional effort. Their encouragement from time to time has been an inspiration for everyone.

The publisher and the editorial board hope that this book will prove to be a valuable piece of knowledge for researchers, students, practitioners and scholars across the globe.

List of Contributors

Goldshleger Naftaly
Civil Engineering Ariel University, Israel
Israel Ministry of Agriculture, Beit Dagan, Israel

Basson Uri
Geosense Ltd., Even Yehuda, Israel

Oleksiy Rubel, Alexander Zemliachenko, Sergey Abramov, Sergey Krivenko, Ruslan Kozhemiakin and Vladimir Lukin
National Aerospace University, Ukraine

Benoit Vozel and Kacem Chehdi
University of Rennes 1, France

Arshad Ashraf, Manshad Rustam, Shaista Ijaz Khan, Muhammad Adnan and Rozina Naz
Climate Change, Alternate Energy and Water Resources Institute (CAEWRI), National Agricultural Research Center, Islamabad, Pakistan

Pasquale Imperatore and Antonio Pepe
Istituto per il Rilevamento Elettromagnetico dell'Ambiente (IREA), National Research Council (CNR) of Italy, Napoli, Italy

José Mauricio Galeana Pizaña, Juan Manuel Núñez Hernández and Nirani Corona Romero
Centro de Investigación en Geografía y Geomática "Ing. Jorge L. Tamayo", México DF, Mexico

Guido Staub and Catherinne Muñoz
Universidad de Concepción, Departamento de Ciencias geodésicas y Geomática, Los Ángeles, Chile

Index